EARTH

CHRIS PACKHAM
&
ANDREW COHEN

EARTH

OVER 4 BILLION YEARS
IN THE MAKING

WILLIAM
COLLINS

William Collins
An imprint of HarperCollins*Publishers*
1 London Bridge Street, London SE1 9GF

WilliamCollinsBooks.com

HarperCollins*Publishers*
Macken House, 39/40 Mayor Street Upper
Dublin 1 D01 C9W8, Ireland

First published in Great Britain by William Collins in 2023

2026 2025 2024 2023
10 9 8 7 6 5 4 3 2 1

ISBN 978-0-00-850720-6

Series design created by Zoë Bather
Design: D & N Publishing, Baydon, Wiltshire
Editorial: Helena Caldon

Printed and bound by GPS Group
in Bosnia and Herzegovina

INTRODUCTION

'You have to know the past to understand the present.'
Carl Sagan

Above: The TW Hydrae protoplanetary disc photographed by the Atacama Large Millimeter Array (ALMA) telescope. The multiple rings and gaps indicate the presence of emerging planets as they sweep their orbits clear of dust and gas.

Opposite: Illustration of our Solar System's formation. The Sun's gravity attracted dust and small particles that gradually became asteroids and eventually planets.

Every great story has a beginning, and the story of the Earth is no different. It's a tale that began over 4.5 billion years ago, when this 6-billion-trillion-tonne ball of beautiful, life-sustaining rock began forming out of the dust of the newly developing Solar System, which had a young Sun at its centre and clouds of gas and dust encircling it. The planet you are sitting on right now was, as unbelievable as it seems, slowly pieced together by the collision of endless cosmic scraps over millions upon millions of years. Formed by nothing more than the force of gravity, dust became rock, rock became boulder and boulder became a molten lump so vast it would pull itself into the near-perfect sphere that we know and inhabit today.

But although this was the beginning of our planet's story, it was far from the beginning of a world that we could recognise today. At its birth there was no life, no ever-changing sky, and there were no oceans. This was a planet born out of a series of endless cosmic collisions, and in its early years it was still reeling from the violence of that creation. To understand how it would become not just a planet but a home, we need to weave our way through the events of this young, hostile world's early life, as it hurtled from one catastrophic event to another. At times it seemed as if there would be no life-filled future for Earth, as again and again it faced the prospect of becoming a sterile world, choked by conditions that had become too extreme – whether too hot, too cold or just too toxic – to allow anything to survive.

And yet, despite everything that was thrown at it, life would ultimately conquer and flourish. Our planet has created every living thing – including you and me. Today, we see all of this beauty, but we also take it for granted. This is a living, breathing, life-sustaining planet that is more fragile than we like to admit and more indifferent to us than we want to acknowledge.

The story of our planet is long, and we are just beginning to understand the extraordinary moments, the endless twists and turns and, to be blunt, the incomprehensible luck that has led to all of this. This is because the Earth's transformation was not an inevitability – in fact, it wasn't even likely.

This is the almost implausible story of our Earth; how it went from a fiery, dead planet to a beautiful, living, breathing, blue bubble floating in the darkness of space.

GEOLOGIC TIMELINE OF EARTH'S HISTORY

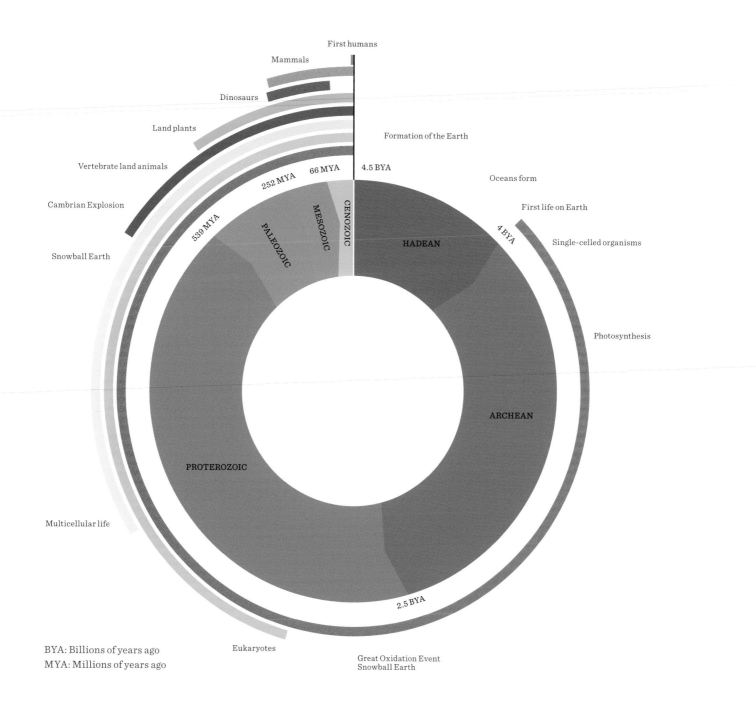

First humans

Mammals

Dinosaurs

Land plants

Vertebrate land animals

Cambrian Explosion

Snowball Earth

Multicellular life

Eukaryotes

Formation of the Earth

Oceans form

First life on Earth

Single-celled organisms

Photosynthesis

Great Oxidation Event
Snowball Earth

252 MYA   66 MYA   4.5 BYA

539 MYA

4 BYA

2.5 BYA

HADEAN

ARCHEAN

PROTEROZOIC

PALEOZOIC

MESOZOIC

CENOZOIC

BYA: Billions of years ago
MYA: Millions of years ago

GEOLOGIC TIME SCALE

EONS	ERAS	PERIODS	SUBPERIODS OR EPOCHS	
PHANEROZOIC	CENOZOIC	QUATERNARY	HOLOCENE	11,700 years ago
			PLEISTOCENE	
				2.58 MYA
		NEOGENE	PLIOCENE	5.33 MYA
			MIOCENE	
				23.03 MYA
		PALEOGENE	OLIGOCENE	33.9 MYA
			EOCENE	
				56 MYA
			PALEOCENE	66 MYA
	MESOZOIC	CRETACEOUS		
				145 MYA
		JURASSIC		
				201 MYA
		TRIASSIC		
				252 MYA
	PALEOZOIC	PERMIAN		
				299 MYA
		CARBONIFEROUS	Pennsylvanian Subperiod	
			Mississippian Subperiod	
				359 MYA
		DEVONIAN		
				419 MYA
		SILURIAN		444 MYA
		ORDOVICIAN		
				485 MYA
		CAMBRIAN		
				539 MYA
PROTEROZOIC	NEOPROTEROZOIC	EDIACARAN		635 MYA
				2.5 BYA
ARCHEAN				
				4 BYA
HADEAN				
				4.5 BYA

Life on Earth is a glorious abundance, richly diverse and phenomenally complex and interconnected. Our seas sparkle with fish species like these sardines, our skies ring with the cries of birds and our forests are shaded by towering trees. It's worth taking a moment to appreciate how fantastic and how unlikely all this life really is, and to consider how we can protect and support it for all our futures.

ATMOSPHERE

'Understanding how Earth got
its atmosphere is the first step in
understanding how to protect it.'
Professor Peter Girguis, Harvard University

THE JAWS OF HELL

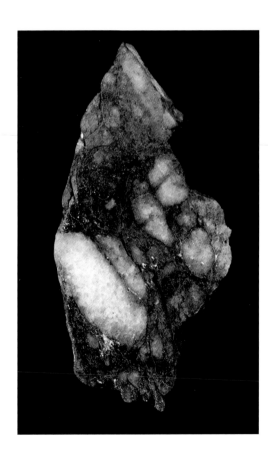

We begin our story with a young, unrecognisable and alien planet. Just a few million years after its birth, this is a volatile world, still burning with the residual heat of its formation and under almost constant bombardment from a chaotic, asteroid-filled solar system. In a sky that is black both by day and night, there is no atmosphere to diffuse the Sun's rays, because space meets Earth directly at its surface. The Sun burns in this inky sky as a ball of brilliant white light, while a newly formed moon hangs heavy, still molten from its own formation. The lunar presence sits tucked in close to its parent planet, illuminating a landscape that is hellish in every direction.

This is the Hadean Eon, a period of Precambrian time named after the Greek god of the Underworld, the first and perhaps the most mysterious of Earth's geological time frames. The Hadean began with the formation of the planet 4.5 billion years ago and ended over half a billion years later. Little direct evidence remains of this vast era of time, because the constant bombardment of the Earth's crust by meteorites, combined with the dynamism of the planet's ever-changing exterior, has long since destroyed or hidden any trace of its ancient surface. Only the rarest of fragments remain for us to examine and ponder what that ancient surface might have looked like.

The oldest of all of these, in fact the oldest piece of the Earth we have ever discovered, was found in the Jack Hills of the Narryer Gneiss Terrane, in an area of Western Australia known for its ancient rock formations. This is no majestic slab of ancient rock, though; it is the merest trace of zircon, a crystal that is found in tiny amounts in almost all of the Earth's igneous (volcanic) rocks. Zircon is an incredibly hard, durable and chemically inert mineral, and for that reason it can survive for immense periods of time. It also contains trace elements of uranium and thorium, whose slow but predictable radioactive decay can provide the perfect method for accurately measuring the age of a rock sample. Through these radiometric dating techniques we've been able to put an age on samples of zircon from the Jack Hills that date back over 4.4 billion years, right to the middle of the Hadean. This is the oldest-known material of any kind formed on our planet, and it gives us a window into that alien world around 100 million years or so after its birth, to a time when the Earth was cooling just enough for this rock to be perhaps part of that first crust, the first solid planet, and a stage upon which the greatest of stories could begin to play out.

Today our world couldn't be more different. It's a place of vivid colour, where oceans give way to land, creating myriad environments and habitats filled with a seemingly endless array of life forms. And yet all that divides our living world from the uninhabitable conditions of the cosmos beyond is gossamer-thin layers of gas stretching upwards by barely over 100 kilometres. This atmosphere is our great protector, a thin blue line that nurtures all life beneath it and marks out our planet as perhaps unique in the known Universe.

In this chapter we will tell the almost implausible story of how this great protector came to be, how it was born out of a turbulent, explosive young planet and how it created the setting for a water world to take shape. We will also explain how these events allowed a set of conditions to form where life could emerge and flourish, and begin an intimate four-billion-year dance between our atmosphere and life itself, creating, shaping and calibrating that atmosphere in ways that we are only just beginning to fully understand.

Above: The oldest-known material of terrestrial origin: zircon 4.4 billion years old found in the Jack Hills, Western Australia.

Opposite: These iron-rich rocks in northwestern Australia formed before the presence of atmospheric oxygen, and life itself.

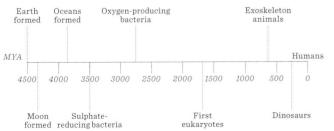

Earth formed   Oceans formed   Oxygen-producing bacteria   Exoskeleton animals

MYA   Humans

4500   *4000*   *3500*   *3000*   *2500*   *2000*   *1500*   *1000*   *500*   *0*

Moon formed   Sulphate-reducing bacteria   First eukaryotes   Dinosaurs   A TIMELINE OF THE EARTH

A bird comes into view; its feathers are silvery, rippling in the wind as it glides along the edge of this escarpment.

With a wingspan of more than 3 metres, the Andean condor weighs up to 15 kilograms, so in terms of their weight and wingspan these are the largest flying birds on Earth.

The condor launches itself out into the air to catch a thermal, and in this way it can travel hundreds of kilometres without hardly ever beating its wings. Watching these giant Andean condors soaring here reveals how their life is completely intertwined with that thin cloak of air that is wrapped around our planet. But when you think about it, everything – every plant, every fungus, every bacterium, every tiny insect, every giant reptile, even every human being – is completely dependent on this atmosphere.

A THIN BLUE LINE

Let's start by looking up. Wherever you are on the planet at this very moment, you, like every single one of us, are swimming in an ocean of air that is as ethereal as it is majestic. Stretching 100 kilometres above you, this is an ocean made of nitrogen (78.08 per cent), oxygen (20.95 per cent), argon (0.93 per cent), carbon dioxide (0.038 per cent) and a collection of other gases in trace amounts. Held to the surface of the planet by nothing more than gravity, this protective blanket that we collectively call air has an overwhelming effect on the characteristics of our planet and all life upon it. Without the pressure created by our atmosphere there would be no liquid water for life to exist in, ultraviolet solar radiation would bombard the surface, the temperature would vary dramatically between day and night, and the whole planet wouldn't be able to hold onto the heat that, through a process we know as the greenhouse effect, helps maintain a stable, life-supporting climate. It is the air that we inhale, filling our cells with the combustible oxygen we need to drive every process in our body. It also provides carbon, the crucial ingredient for all photosynthesising plants and the foundation of almost every food chain on Earth. This is the basis of our weather and the reason we live under the most brilliant blue skies and experience the multitude of colours that usher the Sun in and out of our lives each day.

But our atmosphere is not one homogeneous entity or a single layer of nurturing gas, it is a complex and dynamic system that varies significantly with altitude. From the surface of the Earth upwards, the general rule is that the air pressure and density decrease consistently as you ascend. However, the temperature of the atmosphere does not follow such a straight trajectory and may remain constant or even increase with the changing altitude. We've plotted this temperature signature as you climb into the atmosphere and away from Earth in great detail, using

EARTH'S ATMOSPHERE

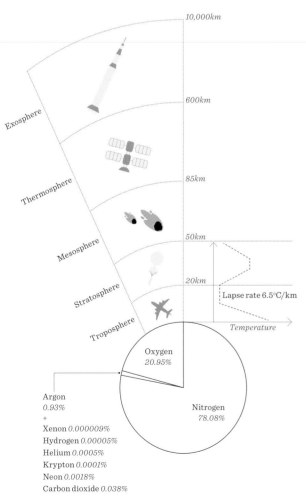

10,000km

600km

85km

50km

20km

Exosphere

Thermosphere

Mesosphere

Stratosphere

Troposphere

Lapse rate 6.5°C/km

Temperature

Oxygen
20.95%

Argon
0.93%
+
Xenon 0.000009%
Hydrogen 0.00005%
Helium 0.0005%
Krypton 0.0001%
Neon 0.0018%
Carbon dioxide 0.038%

Nitrogen
78.08%

'Our atmospheric composition may actually be the thing that is screaming out to the cosmos, a signal to the Universe that we're here, that life exists on our world.'
Dr Michael Wong, Carnegie Institution for Science

weather balloons containing ever-more-advanced instrumentation. Known as the lapse rate, this rate of temperature change with altitude is used to demarcate the structure of the Earth's atmosphere into five distinct layers.

The first of these is the troposphere, which extends from the surface of the Earth up to an average height of 12 kilometres – being lowest at the poles and highest at the equator. This layer contains the vast majority of the mass of the atmosphere (up to 80 per cent), even though it is only a thin slither sitting beneath another 10,000 kilometres of air. In fact, 50 per cent of the total mass of the Earth's atmosphere sits in just the first 5.5 kilometres. The troposphere contains nearly every living thing on the planet, supporting every plant with the carbon dioxide needed for photosynthesis and every animal with the oxygen needed for life. It also contains 99 per cent of all the water vapour on the planet, which means this is where the weather happens, because almost every cloud in the sky sits in the troposphere, except for the tips of cumulonimbus thunderclouds whose summits can peek up into the level above.

Every day, the heat of the Sun that has been absorbed radiates out from the surface of the Earth and up into the troposphere, which means that the lowest parts of the troposphere are the warmest and the highest parts the coolest, creating the powerful convection currents that help drive our weather.

It's this gradual reduction in temperature with altitude that demarcates the troposphere from the atmospheric layer that sits directly above it – the stratosphere. At the boundary of these two distinct layers, around 12 kilometres above the surface of the Earth, is the tropopause, the place in the planet's atmosphere where the temperature stops dropping with altitude and instead stays stable or inverts with a layer of warm air sitting above a cooler one.

Opposite: The Explorer II was a manned balloon launched on 11 November 1935 to study the stratosphere. It reached a record altitude of 22,066 metres and carried a two-man crew inside a sealed spherical gondola.

Right: Meteorologists from the Severe Thunderstorm Electrification and Precipitation Study (STEPS) prepare to launch a weather balloon into a storm to gather data on temperature, pressure, wind speed and electrical fields within the thunderstorm itself.

'If we don't understand the history of the atmosphere, how can we possibly be the stewards of it moving forward?'
Dr Lynn Rothschild, NASA Ames Research Center

It's this change in the lapse rate that signals we are now entering the stratosphere, the second-lowest atmospheric layer, which extends from around 12 kilometres above the surface of the Earth to a maximum altitude of approximately 55 kilometres. The stratosphere is defined by a temperature signature totally unlike that of the troposphere – here the temperature rises with increased altitude, in complete contrast to the layer below. Although at first this may seem counterintuitive, the reason for this sudden change is in fact down to a single property of the stratosphere – the ozone layer.

The ozone layer plays an important role in protecting all life on the surface of the planet, including us, so when, in 1974, scientists raised concern about a hole that had appeared in it due to our use of CFCs (Chlorofluorocarbons – a chemical used in aerosols), it was essential to act quickly to repair the damage to this crucial part of our atmosphere. (An interesting example of humans actually acting on the evidence when faced with a growing atmospheric threat!)

To understand the importance of the ozone layer you have to understand the chemistry of its key singular ingredient. Ozone is what is known as an allotrope of oxygen, a structurally different form of the gas with the chemical formula O_3, as opposed to O_2. Just as allotropes of carbon, such as diamond and graphite, have strikingly different physical properties, ozone differs fundamentally from the O_2 that fills the troposphere, most significantly in its relationship with ultraviolet light raining down from the Sun. Found mainly in the lower levels of the stratosphere, at a height of between 20 and 40 kilometres, the ozone layer absorbs huge amounts of the Sun's ultraviolet radiation – over 97 per cent of its medium-frequency ultraviolet light. This is the characteristic that makes this layer so important in protecting the life forms beneath it from the potentially damaging effects of UV exposure, particularly UV-C exposure, which is harmful to almost all living things because of its ability to cause genetic damage.

The absorption of all this UV light by ozone molecules in turn triggers a chemical reaction, whereby the ozone is photolysed (broken down) by the UV into O and O_2, which rapidly re-forms into ozone and results in the release of heat. It's this process of photolysis (the separation of molecules by the action of light) and rapid reformation of ozone that creates the characteristic temperature inversion of the stratosphere. The heat generated by this process causes temperatures to rise from around -51 degrees Celsius at the base of the stratosphere to as much as 0 degrees Celsius at its highest levels. With very little mixing between the different temperature levels due to the lack of convection currents, this layer is *stratified* into layers of warmer air sitting above cooler layers below, and it is for this vertical stratification that it is named. This process also gives the layer a degree of stability; in the absence of convection currents and the associated weather systems that are constantly driven through the troposphere, there is hardly any turbulence in the stratosphere. This makes the lower levels the ideal altitude for commercial airliners to cruise in, because the lower density of the air combined with the lower temperature means this flight path is the most fuel-efficient. Beyond these reaches of the stratosphere the air becomes too thin to support commercial flight, so we leave behind the last layer of the Earth's atmosphere regularly populated by humans and make our way upwards towards the next layer, known as the mesosphere.

The third layer of our atmosphere, the mesosphere, rises up from an altitude of around 50 kilometres and extends up to 85 kilometres above sea level. Here, once

Opposite: In strong sunlight, the hippopotamus sweats a red substance that blocks out UV rays and thus acts as a natural sunscreen.

Above: A university student smells a magnolia through a gas mask on the first Earth Day in 1970, protesting the pollution that has damaged the ozone layer.

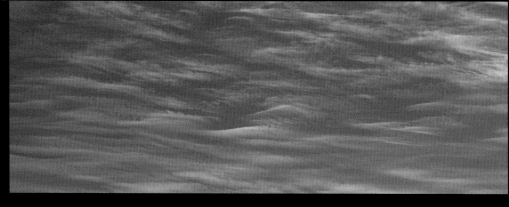

Cirrostratus clouds are
similar to cirrus, but form
more of a widespread,
veil-like layer.

5km

Cumulonimbus clouds are
menacing-looking multi-
level clouds, extending high
into the sky in towers or
plumes. More commonly
known as thunderclouds,
cumulonimbus is the only
cloud type that can produce
hail, thunder and lightning.

2.5km

Cumulus are fluffy,
fair-weather clouds that
sometimes look like pieces
of floating cotton wool.
The base of each cloud is
often flat and may be only
1,000m above the ground.
These clouds grow upward,
and they can develop into
cumulonimbus.

Stratus clouds are uniform
and flat, producing a grey
layer of cloud cover which
may be precipitation-free
or may cause periods of light
precipitation or drizzle.

Cirrus comes from the Latin word meaning 'curl of hair'. These are high clouds, forming between 6.2 and 13.7km. They are made up of ice crystals and tend to appear before a low-pressure area like a storm system in the middle latitudes or a tropical system like a hurricane. Their delicate, feathery shape comes from wind currents that twist and spread the ice crystals into strands.

Cirrocumulus clouds are thin cloud patches found high in the troposphere and are made up of individual 'cloudlets'. They appear between 5 and 12km and contain a small amount of liquid water droplets as well as ice crystals.

Altostratus is a middle-altitude cloud genus made up of water droplets, ice crystals, or a mixture of the two. Altostratus clouds are formed when large masses of warm, moist air rise, causing water vapour to condense.

Altocumulus clouds are typically found in groups clumped together. They're found in the middle layer of the troposphere, lower than cirrocumulus and higher than their cumulus and stratocumulus counterparts. The term mackerel sky is also common to altocumulus clouds that display a pattern resembling fish scales.

Stratocumulus clouds are a mix of stratus and cumulus, forming a low, fluffy layer. They may indicate storms to come.

Nimbostratus clouds are multi-level, amorphous, nearly uniform and often dark grey, and they usually produce continuous rain, snow or sleet but no lightning or thunder.

'The evolution of our atmosphere is in many respects the story of the evolution of life on our planet. Life can change a planet fundamentally. It is always this cause-and-effect kind of dance between the environment changing life and life changing the environment.'
Professor Tim Lyons, University of California, Riverside

Above: Noctilucent clouds are very thin and form at altitudes of 75 to 90km. Because they are so thin and high, they can only be seen at twilight, and usually only near the polar regions. These were photographed in Finland.

again, the temperature dynamic follows a simple relationship between increasing altitude and decreasing temperature. This means that by the time we reach the top of the mesosphere, at a region known as the mesopause, we find ourselves at the coldest place around Earth, with an average temperature of -85 degrees Celsius. The air is so cold at this altitude that any water vapour forms into what are known as noctilucent clouds, or, more poetically, *night shining clouds,* around the polar regions of the planet. These are the highest clouds in our atmosphere and if you are lucky enough to be at the right latitude at the right time, when the Sun is rising or setting, this rare phenomenon can be visible to the naked eye.

Devoid of life, both human and otherwise, the mesosphere sits in its lofty position looking down on our planet. A protector of sorts, here the atmosphere first becomes thick enough to slow and then incinerate the millions of meteors that enter Earth's atmosphere each year: protecting the land below from the onslaught of the on-average 40,000 tonnes of rock that rain down on the planet every year.

Together with the stratosphere, the mesosphere makes up what is known as the middle atmosphere of Earth. Beyond its cold higher reaches we enter the upper atmosphere and the second-highest layer of the Earth's atmosphere, a cloudless region known as the thermosphere. For the last twenty-two years this lofty position has been home to, on average, seven human beings living on board the International Space Station. Orbiting at an altitude of about 400 kilometres, the ISS sits in the upper reaches of this atmospheric layer, which extends from a height of 80 kilometres up to a range of between 500 and 1,000 kilometres above sea level. Buffeted by the solar wind raining down on it, the height of the thermosphere ranges so widely due to the ever-changing levels of solar activity, and it is this activity that also creates what is known as the ionosphere in its lower reaches.

This region of the atmosphere is where the sparsely distributed atoms are stripped of electrons, creating in the process perhaps the most beautiful light shows on Earth – the aurora borealis and aurora australis. Once again in this region we find the temperature signature flips, gradually increasing with height to peak as high as 1,500 degrees Celsius. This may be a vital part of the atmosphere, but if we were directly exposed to this atmosphere we would hardly recognise or be able to sense it. The air is so thin it has been calculated that an individual molecule of oxygen would have to travel over 1 kilometre before it would collide with any molecule of air.

Finally, as we reach the upper limits of the thermosphere, at a boundary layer known as the thermopause, we enter the outermost atmospheric layer of Earth – the exosphere. Extending from around 700 kilometres up to 10,000 kilometres above the surface of the planet, this layer ultimately merges with the solar wind. The density of this atmosphere is so low it can't even be described as a gas; instead, it is a collection of atoms and molecules such as hydrogen and helium, and at its lowest point heavier molecules such as nitrogen, oxygen and carbon dioxide. These atoms and molecules are so far apart that they can move across hundreds of kilometres of the exosphere without colliding with anything else, a characteristic that means it can be considered to no longer behave like a gas.

The lower boundary of the exosphere is called the exobase, also known as the 'critical altitude'. There is no measurable air pressure, no change in temperature, and in effect from here on the Earth's atmosphere dissolves into space as the gravitational grip of the planet slowly releases its hold on the most distant particles. At the upper reaches, almost halfway to the Moon, the pull of the Earth that holds our great protector in place and all the life below it finally relinquishes its grip and the atmosphere is no more.

This page: Examples of the trauma of meteor strikes on the Moon (top left) and on Earth, in Western Australia (top right) and in Arizona, USA (bottom).

REDDENING SKIES

Four and a half billion years ago this complex protective structure above our heads didn't exist. On early Earth there was no life, no blue sky, no atmosphere. Apart from the merest wisps of hydrogen and a trace of other gases clinging to the newly formed planet, the void of space reached all the way down to the Earth. So how did all this change occur and from where did the richness of our atmosphere materialise? To answer these questions we don't have to look up but down, because after a hundred million years or so of relative tranquillity it was activity deep within our planet that would transform the conditions on the surface.

Driven by the heat from its formation that had been locked deep within its core 4.4 billion years ago, the Earth was about to explode in a frenzy of volcanic activity. Across the globe molten magma raced up from within, creating rivers of liquid fire across the planet's surface that not only transformed the terrestrial landscape but the atmospheric landscape too, releasing a cocktail of gases that would begin to form the building blocks of our first recognisable atmosphere. This was, however, no place where humans could breathe deeply; consisting mainly of nitrogen and carbon dioxide, this was an atmosphere unlike anything we can see today.

Opposite: Tungurahua volcano in Ecuador releasing plumes of ash, steam and smoke into the air.

Below: Radar image from the space shuttle Endeavour of the Kliuchevskoi volcano in Kamchatka, Russia.

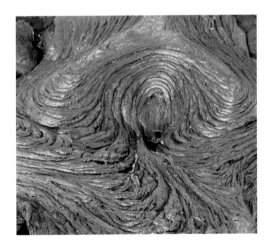

Left: A carbonaceous chondrite meteorite found in northwest Africa.

Above middle: Lava flowing from the active Kilauea volcano in Hawaii, USA.

We know this because we have been able to study the very substance that gave rise to this first atmosphere by analysing one of the rarest types of meteorites that has fallen to Earth. Carbonaceous chondrite meteorites are, in the simplest terms, the very stuff that first built the Solar System. Formed from the dust cloud that circled the newly born Sun over four and a half billion years ago, these ancient artefacts are almost unaltered by time, and so they allow us direct access to the materials that went on to form this planet. It was rocks exactly like this that through a process known as accretion became bigger and bigger, as gravity pulled them into larger clumps until ultimately, four and a half million years ago, trillions of tonnes of rocks and meteorites like this came together. Exposed to the vast pressures inside the newly formed Earth, some of these rocks would have become liquid and molten beneath the surface, before they burst onto the surface and released a cocktail of gases that would form that first thick atmosphere. This is why carbonaceous chondrite meteorites are so valuable; by analysing the exact chemical signature of these meteorites we can model precisely what cocktail of gases this rock would have emitted as it was heated up and exhaled its contents into that early atmosphere.

This type of analysis reveals that these meteorites contain particularly unusual and highly specific versions of chemical elements, including carbon, hydrogen and sulphur, compounds like methane and ammonia, as well as vast amounts of water vapour. Even today these same elements are emerging as gases across the planet from the furnace-like conditions of the inner Earth. So meteorites like this weren't just the building blocks of the planet, they contained the essential ingredients for its atmosphere, and 4.4 billion years ago that had begun to change everything.

Across Earth's surface a reddening shroud gradually began to thicken as the planet's core spilled out its innards in the endless streams of lava that chased across its surface. Over a period of millions of years the vast sea of stars shining down on the Earth's surface slowly started to vanish, until eventually, around 4.3 billion years ago, a new type of dawn began to break across our planet. As the Sun rose, lighting up the barren landscape, Earth's first ever colour-filled sky emerged and a thick haze of gases began to blanket the world. Earth was finally gaining its first substantial atmosphere.

But this ancient sunrise would have looked completely different to the golden ones we see today. Sunlight passing through a thick, churning mixture of methane and carbon dioxide would have given the whole planet a distinctive orange hue. Crucially, this toxic atmosphere was the first substantial shield against space, but to us, Earth would still have been an alien world. Not just because of the noxious orange haze in the sky, or the searing heat on the ground of dark volcanic rock; there was something else fundamentally missing on this early Earth that we take for granted today. Something that would only arrive with the help of this first atmosphere.

Iceland is one of the most volcanically active places on the planet. It is geologically very young – less than 33 million years old – with active plate tectonics and glacial movement as well as volcanoes, and is one of the few places in the world where a divergent plate boundary is exposed at the Earth's surface. The result is this stark, brutal and yet beautiful landscape, which is a good approximation of what the surface of the Earth may have looked like 4 billion years ago.

DAWN OF OUR BLUE PLANET

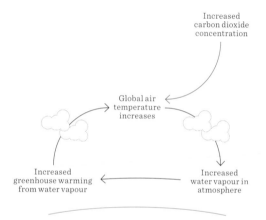

WATER VAPOUR AND THE GREENHOUSE EFFECT
This diagram shows the role of water vapour in climate change on Earth today. As carbon dioxide increases in our atmosphere, due to emissions and loss of plant coverage, air temperatures rise. Evaporation increases and, since warmer air holds more water, water vapour levels in the atmosphere rise as well, which adds to the greenhouse effect, warming the atmosphere further. The cycle reinforces itself.

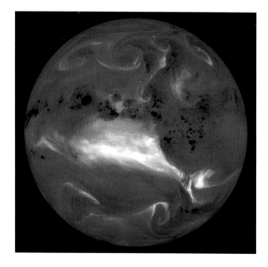

Today, 70 per cent of our world is covered in water. Ours is a planet of almost limitless blue, with endless oceans, rivers and frozen ice caps. All of this water is on a continual journey; it emerges from the leaves of green plants as vapour, then rises up into the clouds where it condenses and falls as rain onto the land, before draining into the rivers that flow into our vast oceans.

The early Earth was a barren planet. For the first half a billion years of its existence there was not a single drop of liquid water anywhere on its surface – beneath the orange haze that smothered the planet, the land was completely dry. But in fact Earth did possess water, and vast amounts of it, in great oceans hidden away not on the surface but in the sky. As the endless volcanic activity continued to pump a cocktail of toxic gases into the newly forming atmosphere, it carried with it enormous quantities of water in vapour form. Lifted into the air of this darkening atmosphere, but with temperatures consistently high, that was where the water stayed, trapped above the baking ground and seemingly destined to remain as a gas forever.

But as we will see in the story of the Earth, nothing stays the same for long. Slowly, across millions of years, the intense heat left over from its formation that was trapped within its core gently lessened, drifting back out into space. The planet was slowly cooling, its surface hardening, and with it the conditions in that primordial atmosphere began to change. As the atmosphere cooled, it allowed the very highest water vapour to start condensing out of the air. Imperceptibly at first, high up in the atmosphere minute droplets, just a fraction of a millimetre across, were beginning to form. At first these were so light and small they just floated, soaring on moving air, but gradually, as the conditions changed and these minuscule droplets began colliding and merging, they eventually reached a tipping point. Earth's gravity began to draw them down towards the ground, and for the first time water droplets – rain – began to fall through the atmosphere. But this journey was short-lived. Just as we see on Earth today in the troposphere, the closer to the ground you get, the warmer it becomes, and on our early planet this situation was even more extreme. Temperatures in the lower atmosphere were fiercely hot and this searing heat meant not a single drop of rain could make it to the surface, evaporating back into water vapour long before it made groundfall.

It would take more and more heat to leak out from the planet into space to allow the atmosphere to become cooler and enable the drops of water to get closer and closer to the surface. Eventually, though, on one particular day in our planet's history, at one particular place, at one particular moment, it became cool enough for one of those drops of water to make its way right through the heavens to splash onto Earth's scorched crust. For the very first time in its history, rain had arrived on our planet, and this was just the beginning. Those first few drops were soon followed by the greatest deluge the Solar System has ever seen. The more it rained, the more heat was sucked from the surface and the cooler the planet became, allowing not just more rain to fall but to now settle, transforming our world into something much more recognisable. Earth was becoming a world where huge weather systems swept across the planet and storms sometimes lasting centuries dumped oceans of water from the skies onto the surface. In doing so, this created not just a water-rich atmosphere but a whole new water world, and with it a key element needed for the foundation of life had been well and truly unleashed.

Left: Atmospheric water vapour surrounds the Earth today.

Opposite: Storm clouds laden with water vapour hover menacingly above the horizon.

ONE STRANGE ROCK

Take a look around the Solar System today and you will quickly see that when it comes to atmospheres the Earth is very much an outlier. Of the four terrestrial planets in the inner Solar System – Mercury, Venus, Earth and Mars – only our planet has liquid water on its surface, protected by that shimmering blue veil. Closest to the Sun, Mercury was stripped of any real atmosphere long ago. Battered by the solar wind, it never would have been able to retain a substantial atmosphere, and its small size and weak gravitational pull only add to its limited ability to hold on to the extremely thin collection of gases that surround it today.

Next out, Venus couldn't be more different. Cloaked in a thick choking atmosphere comprising of over 96 per cent carbon dioxide, Venus is nothing like Earth. With a runaway greenhouse effect there is no liquid water, just surface temperatures approaching 500 degrees Celsius and crushing pressures not far short of 100 times greater than here on Earth. It wasn't always this way, though. In its infancy Venus was almost certainly a water world with an atmosphere not unlike the orange shroud of early Earth, but Venus would take a very different course. Rising carbon dioxide levels from its intense volcanism combined with myriad other factors drove the temperature of the atmosphere upwards, eventually evaporating the oceans and driving the greenhouse effect to a point of no return.

Then there is Mars, our red neighbour, which clings onto a cold, thin atmosphere of carbon dioxide and nitrogen, with traces of other gases, including water vapour. Such a thin atmosphere leaves the planet prone to far greater temperature fluxes than Earth, with surface temperatures falling as low as -140 degrees Celsius in winter and as high as 20 degrees Celsius in the summer. Just like Venus, Mars was also far more Earth-like in its youth, surrounded by a thicker, warmer, wetter atmosphere. This was because the greenhouse effect generated by this carbon-dioxide-rich atmosphere allowed temperatures and pressures to rise to a point where liquid water didn't just exist but covered the surface of the red planet with

Below: NASA's MESSENGER spacecraft orbited Mercury between 2011 and 2015 to study the barren, rocky innermost planet.

Below right: Venus, photographed by NASA's Mariner 10 in 1974, is a world of intense heat, crushing atmospheric pressure and clouds of corrosive acid.

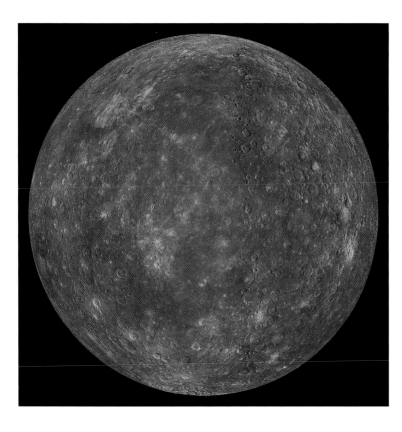

SURFACE TEMPERATURE VERSUS HEATING FACTOR

A comparison of the temperatures and heating factors (essentially, the ability of the atmosphere to hold onto and amplify heat) of the four terrestrial planets of our solar system. Venus, Earth and Mars are isothermal, meaning their temperatures are relatively steady, while Mercury is tidally locked to the Sun, so one side is eternally hot and the other frigid and only 10% of the heat is globally redistributed. The blue band shows where surface liquid water is supported.

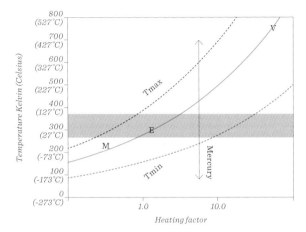

vast seas and river systems whose mark can still be seen today. But once again this wasn't to last. Significantly smaller than Earth, Mars could not hold on to its rich atmosphere, and, stripped away by the solar wind, it too would become a dry and barren land.

Three planets all with hopeful beginnings, but only the Earth has maintained a balance whereby the atmosphere is thick enough to warm the surface and enable liquid water to flow but not be so thick that it chokes the planet dry. The reasons for this delicate balance are numerous and interlinked. The size of the planet, its distance from the Sun, its levels of volcanism, the dynamism of its tectonics and its ability to remove and recycle carbon dioxide from the atmosphere are just a few of the factors that have conspired to keep our atmospheric stability. But for all of that constancy we also know that our atmosphere has changed dramatically over the lifetime of the planet, evolving from the orange shroud that was born out of the Hadean Eon into the oxygen-rich atmosphere we see today. And driving that change more than anything else has been the one force that, as far as we know, has existed and flourished on only one planet in the Solar System.

Life has shaped the destiny of our planet, as much as the planet has shaped the destiny of life. We see this today in the effect that this supposedly intelligent life form is having on the atmosphere of our planet, but it's not the first time life has shaped the atmosphere – that's something that has been going on for billions of years. And yet, if we go back to the very beginning it may well have seemed that life would have been nothing more than a peripheral player in the grand story of our planet.

Below: The Viking I and II probes captured Mars with some wispy cloud cover and polar caps of frozen carbon dioxide and water ice.

Below right: A 2019 satellite image shows the Earth as the uniquely verdant jewel of the inner Solar System.

GENESIS

One hundred million years after oceans formed from the water droplets that had begun to rain down on its surface, the conditions on Earth were transformed. This was now a planet no longer directly exposed to space; its temperatures had become more stable, there was an atmosphere shielding the surface from the Sun's most violent rays, and liquid water on its surface had created a mixing bowl that was allowing the raw ingredients of the Earth to interact in myriad new ways. A new chemistry set using water as its medium was now underway.

What happened next is without doubt one of the great mysteries of Earth's story. What we know for certain is that the sterile ball of rock, riddled with volcanoes and bombarded with meteorites, within the space of half a billion years came alive.

So much about the emergence of life on Earth is unknown. We do not know exactly when it started, where it started, or how, but we do know there was a

Below: Liquid water brought a new era of life and beauty to the Earth.

moment when chemistry became biology and the first 'living' thing came into existence. Whether it was in a shallow pool, around the warming power of a deep sea vent, or perhaps in part delivered to the planet while riding a lift on the back of one of those meteorites, within a billion years of the Earth's birth this was not just a planet but a home.

A little less than four billion years ago, however, this was no Eden, no oasis where life could flourish and prosper. A quarter of a billion years after that moment of genesis, life was just about clinging on around the edges, still fighting to establish itself in this harsh and unforgiving world. With barren black land, acidic green oceans and a deadly orange atmosphere, it was a world that would have been totally uninhabitable for the vast majority of life forms today.

A COMMON ANCESTOR
This diagram outlines every living thing on Earth and, astonishingly, they can be traced back, using genetics, to the point where you can see all life on Earth has come from a single organism. We call it the 'Last Universal Common Ancestor' – LUCA for short.

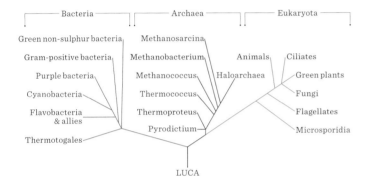

Below left: Deep sea bacteria (blue) on the surface (yellow) of annelid worms (*Alvinella* sp.). These bacteria are chemosynthetic, which means that they produce energy from chemicals (sulphide compounds from deep sea vents) instead of from sunlight. They can withstand temperatures up to 105 degrees Celsius.

Below: A black smoker hydrothermal vent in the Kermadec volcanic island arc, in the South Pacific Ocean, provides an unusual habitat for life, perhaps mimicking some of the earliest conditions on Earth.

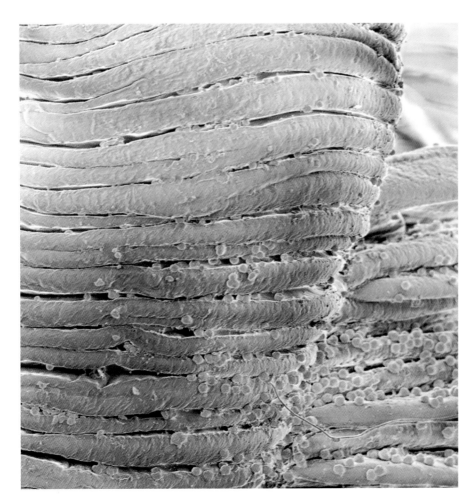

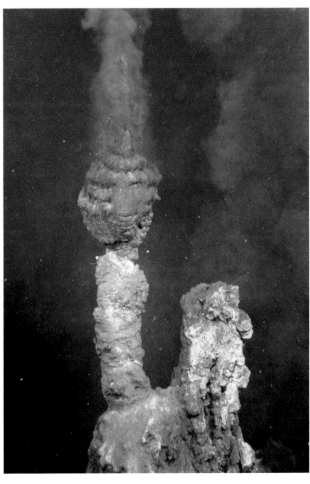

LIFE IN THE EXTREME

Above: Filming *Earth*, the series, on location at the El Tatio geyser field.

Opposite: The geysers at El Tatio geothermal field high in the Andes of northern Chile produce mineral-rich steam.

Above: Arsenic is now known to be incredibly toxic, but in the Victorian era it was popularly used to create vivid green fabrics and wallpapers like those by William Morris.

In a handful of places on the planet we can still catch a glimpse of what life here might have looked like all those millions of years ago. The El Tatio geyser field in the Andes mountains of northern Chile is one such location. The largest geothermal field in the southern hemisphere, covering an area of around 30 square kilometres and at an elevation of over 4,000 metres, El Tatio experiences a wide variety of geothermal activity, including hot springs, geysers, mud volcanoes and boiling water fountains. It provides an endless, unpredictable spectacle that is at its most impressive when you visit in the cool air of the early morning.

But not everything that is special about this place is immediately obvious to those lucky enough to visit. Amidst the superheated water and steam, something extraordinary and microscopic is flourishing within these scalding pools. With the water in the vents reaching 85 degrees Celsius, this is an environment that is easily warm enough to denature the proteins in you and me, or, to put it another way, it's hot enough to cook you! The near-boiling water also contains one of the highest concentrations of arsenic, a deadly toxin, that can be found anywhere in the world.

And yet for all of these foreboding characteristics, the beautiful array of colours around the pools is an indication that this landscape is far from sterile. Life does not only survive here, it flourishes in colourful mats made up of billions of thriving primordial bacteria called extremophiles.

Environments like this were the first to grab the interest of scientists studying the limits of life on Earth. Back in the 1960s this was a peripheral and often frowned-upon area of scientific research; however, a small group of biologists working in hot spring environments like El Tatio began to realise that exploring life in these extremes could reveal fundamental principles about all life on Earth and perhaps the Universe beyond. Among the first such living organisms to be isolated and documented in any detail was a temperature-resistant bacterium named *Thermus aquaticus*, which was discovered living in the hot springs of Yellowstone National Park, in the United States.

Until this discovery scientists thought life, even in hot springs, couldn't survive in temperatures above 55 degrees Celsius, but nobody had really taken the time to study the life in these environments in any detail. Then in 1969 two scientists from Indiana University, Thomas Brock and Hudson Freeze, published their groundbreaking study of *T. aquaticus*. Discovered in a sample they collected from one of Yellowstone's famous terrestrial hot springs, known as the Mushroom Pool, back in 1964, here was an organism that thrived at an optimal temperature of 65 to 70 degrees Celsius and survived in up to 80 degrees Celsius. As Brock said in an interview years later, 'I was stunned by all these microbes that were living in all these hot springs and nobody seemed to know anything about them.'

Intriguingly, *T. aquaticus* has gone down in history not only as the first fully characterised extremophile but also as a key discovery that led to the explosion of DNA technology in the last fifty years. In a beautiful example of the power of pure scientific curiosity, the bacteria that Brock and Freeze discovered ended up providing a heat-resistant enzyme that is one of the most important components used in the Polymerase Chain Reaction – the DNA amplification technique that is behind every aspect of DNA fingerprinting used across so many different fields of biology, medicine and forensics.

Since those early discoveries in the 1960s, research into life forms that live in the most extreme environments on Earth has exploded, with a whole new field of science – named exobiology – trying to answer fundamental questions about life on this planet and beyond. As we have looked across landscapes like the frigid deserts of Mars or the boiling pressure-cooker surface of Venus, we can begin to wonder what, if any, life could survive there. What are the most extreme temperatures,

'The environment at El Tatio geysers is extreme for many reasons. The water is boiling hot and the altitude means the area is subject to a lot of the Sun's ultraviolet radiation. The colours in mats around the pools are billions of thriving bacteria that have carved out a niche. It's incredible to find so much life, not just surviving but thriving.'
Usha Lingappa, University of California, Berkeley

pressure, acidity and toxicity that limit ability for life as we know it to persist? These questions have driven astrobiological research, however, the answers aren't found on distant planets but are found here, in the most extreme environments on Earth, like El Tatio, where there is a great range of species and an enormous number of individual organisms. These include mats of orange and green cyanobacteria, single-cell archaeans like thermoproteota, and in the warmest pools a range of hyperthermophiles – organisms that thrive in temperatures of 60 degrees Celsius.

All of this biology tells us that the environmental boundaries for life are far broader than we once thought, and even the simplest life is inherently flexible in its ability to adapt to what at first can appear to be a toxic environment.

And, of course, they are not just a snapshot of the possibilities of life in the Universe today, they also offer a window into the past. Scientists are drawn to places like El Tatio because they provide an analogue of the environmental conditions of the early Earth and a glimpse of the challenges that those very life forms that existed on the planet almost four billion years ago may have faced.

You see, living in these extreme environments comes with severe limitations because the extremophile bacteria clinging on around these hot springs are essentially locked into this environment, one that is defined by very precise requirements in terms of the heat of the water and the nutrients in it. If we were to remove these life forms from this highly specialised environment they would almost certainly die, and this environmental limitation was pretty much the same for that early life on Earth. It was essentially stuck, trapped in the niches that it evolved to survive in. Early life wasn't prolific, widespread or even visible, it was vulnerable, hidden and stuck, and so for around the first billion years of life on Earth, relatively little changed. Living under a storm-filled orange atmosphere, this simple life was just existing, hidden beneath the waves, and the development of any more complex life was far from an inevitability.

After millions and millions of years of seeming stasis, though, something changed. A single development in a single cell that would alter the way life existed forever. In a quantum leap forward, a radical new type of life evolved, an organism with a new ability that could transform not only the biosphere but the atmosphere as well. The ancestors of those revolutionary early organisms are still around us today – they can be found hidden in almost every puddle, lake, sea or ocean across our planet.

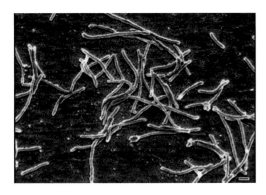

Above: The heat-loving bacteria, *Thermus aquaticus*.

Right: The geysers and pools found in Yellowstone National Park, Wyoming, USA, are excellent sites to study extremophiles.

Opposite top: As we learn more about the adaptability of life on Earth, the other planets in our Solar System seem less unequivocally hostile than was previously assumed. This carbon dioxide ice landscape on Mars is created by vents on the planet's surface belching out the gas.

Opposite bottom: Meanwhile, on Venus the planet's surface nearly boils (here lava extends for hundreds of kilometres around the volcano Sapas Mons), but the atmospheric pressure and temperature at about 50–65km above the surface of the planet is nearly the same as that of the Earth.

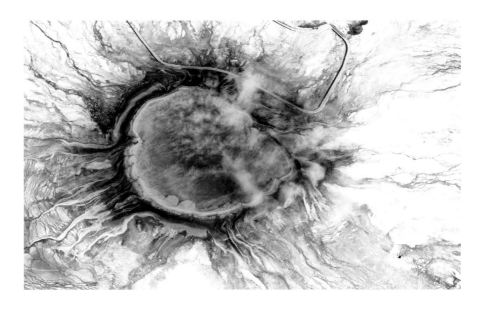

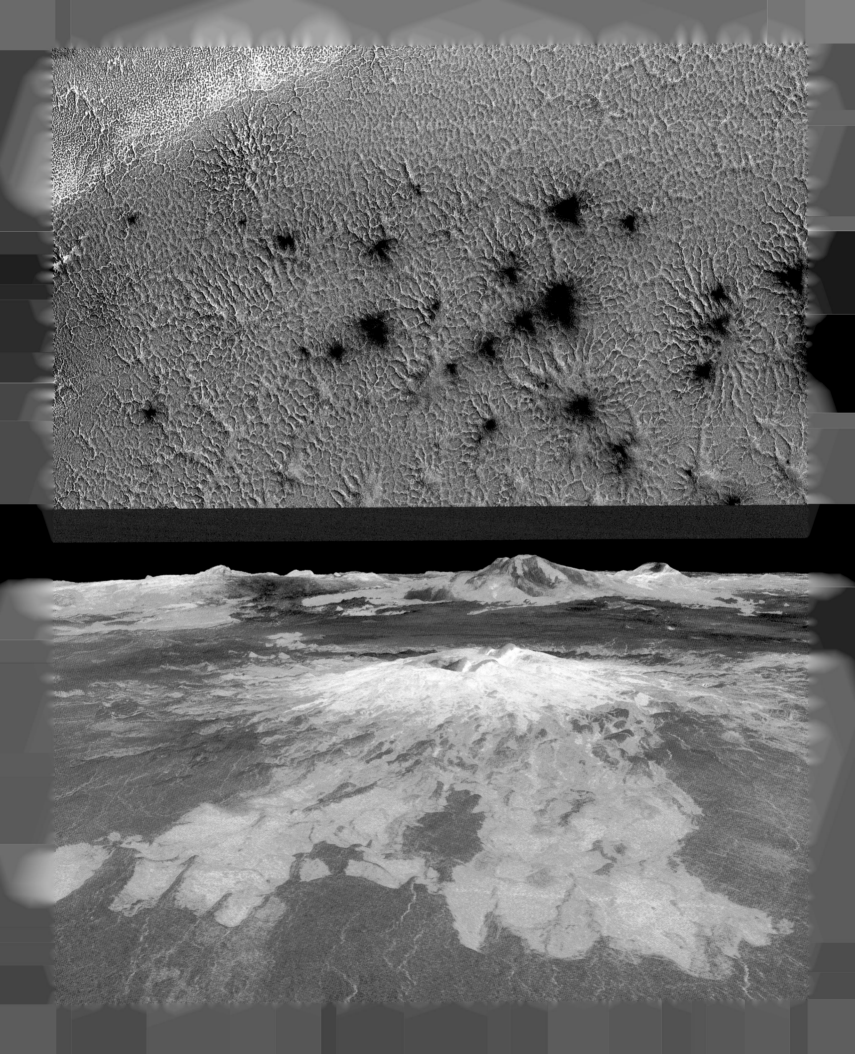

Peering down through this microscope is like taking a look back at life on Earth almost three and a half billion years ago. These rod-like structures here are cyanobacteria, and we think they are pretty similar to those that existed trillions of generations ago, when our atmosphere was very different. But I know what you are thinking: they are not very impressive, are they?

Nevertheless, they are probably one of the most successful organisms ever to live on planet Earth, because a little over three billion years ago these tiny flecks, these microscopic organisms – just a fraction of a millimetre across – started to build an atmosphere in which humans could live and breathe.

This revolutionary process is called photosynthesis.

CATCHING SUNBEAMS

It's time to go back to school and remember one of the simplest, yet most powerful equations in the whole of science:

$$6CO_2 + 6H_2O \Rightarrow C_6H_{12}O_6 + 6O_2$$

This is, of course, the equation for photosynthesis, the process by which plants use the energy of sunlight to spin sugar out of water and carbon dioxide. It's a process that underpins so much of life on Earth, capturing the vast supply of energy that hits the planet as sunlight and turning it into the basic food that is then carried through the complex food web, fuelling endless life forms – including you and me.

The discovery of photosynthesis in the later decades of the eighteenth century is one of those beautiful examples of incremental scientific research; the baton pass of knowledge that ultimately leads to a new fundamental discovery.

It began with Jan Van Helmont, a chemist, physiologist and physician from Brussels who started exploring the growth of plants to see if they were directly gaining mass from the very soil they stood in – a fair assumption in the mid-seventeenth century. Through careful measurement of a willow tree over five years of study, Van Helmont was able to discount the idea that the mass of the soil was transferred to the plant to allow it to grow and instead hypothesised that it was the water he was adding to the potted plant that provided the mass for its development. This was, as we now know, not entirely correct, but it pointed the scientists who followed in a new direction of investigation.

Next up was Joseph Priestley, a chemist and minister who, in 1771, discovered that if you put a mouse in an inverted bell jar the air would become 'injured' and the mouse would soon fall unconscious. But add a plant to the bell jar and the mouse would be revived. It was an intriguing phenomenon without an absolute explanation, but it suggested that the plant was adding something to the air that was crucial to life. In 1779, Jan Ingenhousz repeated Priestley's bell jar experiment but with a twist, revealing that the plant would only revive the mouse if the experiment was conducted in the presence of direct sunlight. Cover the bell jar with a cloth and the mouse was as doomed as before – whether the plant was in proximity or not.

Slowly the ingredients behind the growth of a green plant were being revealed: sunlight, water and the release of a mysterious life-giving substance brought a flailing mouse back from the brink. By the end of the century the basic reaction behind photosynthesis was laid out. In the 1790s Jean Senebier, a Swiss pastor and botanist, demonstrated that green plants use carbon dioxide and release oxygen when exposed to sunlight. And by 1808 Nicolas Théodore de Saussure was able to confirm that it was the combination of water and carbon dioxide that provided the mass for a plant to grow, although it wasn't until 1893 that the term photosynthesis (from the Greek, *photo* – light, and *synthesis* – put together) was proposed by Charles Reid Barnes.

Today we take this understanding of photosynthesis for granted, and over the last century we have added endless layers of complexity in our knowledge of the molecular and chemical pathways that are needed to turn carbon dioxide and water into sugar and oxygen. We live on a planet where we know that our lives depend on the green plants that both feed us and fill our atmosphere with life-giving oxygen. But despite the fundamental process of photosynthesis appearing to be almost universal across the planet, there was a time when the process was dramatically different and far from the green life-support system that we see all around us today.

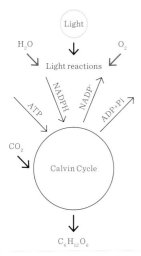

PHOTOSYNTHESIS WITH OXYGEN
Photosynthesis changes sunlight into chemical energy, splits water to liberate O_2, and fixes CO_2 into sugar through the Calvin Cycle.

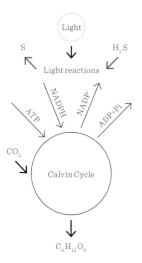

PHOTOSYNTHESIS WITHOUT OXYGEN
Sulphide (H_2S) is used as a reducing agent during photosynthesis in green and sulphur bacteria.
1. The light-dependent reactions occur when the light excites a reaction centre, which donates an electron to another molecule and starts the electron transport chain to produce ATP and NADPH, the 'energy currency' of cells.
2. Once NADPH has been produced, the Calvin Cycle proceeds as in oxygenic photosynthesis, turning CO_2 into sugar.

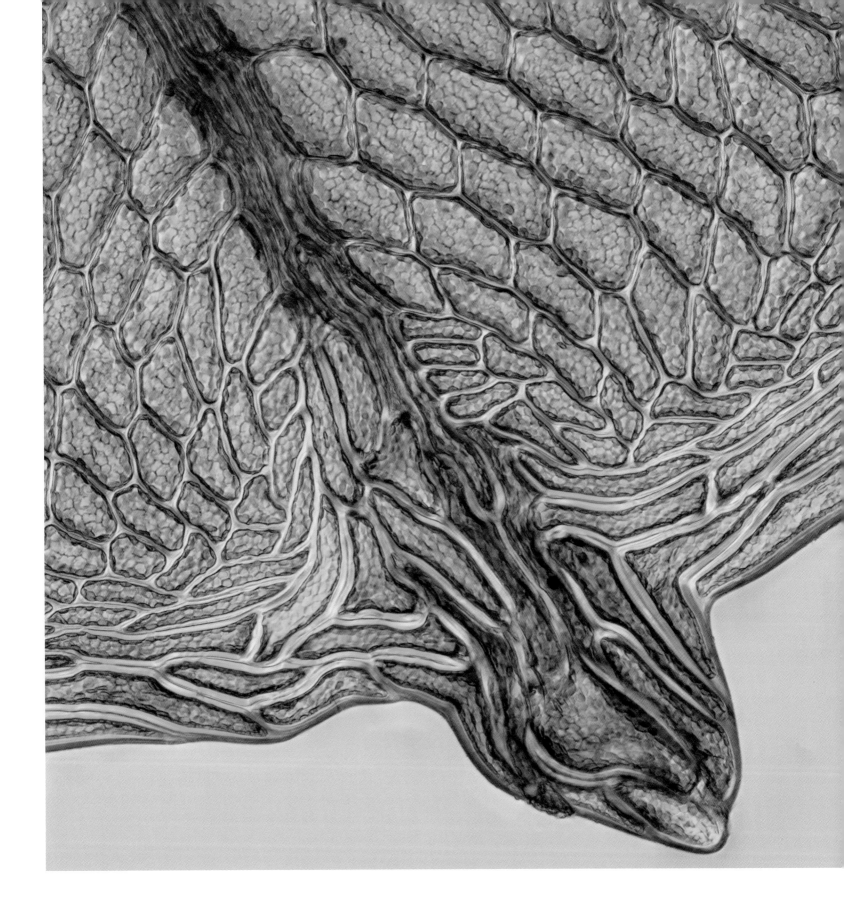

Above: Polarised light micrograph of the underside of a moss (*Cinclidium stygium*) leaf, showing the presence of photosynthesising chlorophyll cells.

We don't know exactly when life first evolved the ability to harvest the energy of sunlight, but fossil evidence suggests that there was life capable of photosynthesis on the planet at least 3.4 billion years ago. But those early photosynthetic organisms were far from the green plants we see today. Bacteria living in the near-toxic oceans and lakes of what is known as the Archean Eon (about 4 billion to 2.5 billion years ago) evolved an ability to use sunlight to create sugars not by using water but by using other chemical ingredients to build the pathways that delivered energy to the organism. This primordial process of photosynthesis still uses sunlight interacting with carbon dioxide, but with a reducing agent like hydrogen sulphide rather than water to donate the electrons needed to drive the creation of the complex sugar molecules. It's a process that results in the production of water as a waste product as well as elemental sulphur, but crucially there is no release of oxygen. This is photosynthesis but not as we commonly understand it, and it is not the oxygen-enriching, life-giving process that we associate with our green planet today.

We still find life forms that use anoxygenic photosynthesis on Earth today, including several groups of bacteria, such as green sulphur bacteria, acidobacteriota and heliobacteria, but they remain in the relative shadows compared to their photosynthesising green relatives. And yet 3.4 billion years ago the situation was very different. For almost a billion years after the first anoxygenic photosynthesis began on our planet, no other type of photosynthesis existed – no water had been torn apart and therefore there was still no oxygen in the atmosphere. But in one of the most important metabolic innovations in Earth's history, that was all about to change.

Today, cyanobacteria are some of the most abundant organisms on the planet; otherwise known as blue-green algae, they are found across almost every environment on Earth. One of the most common species, *Prochlorococcus*, is just 0.5 to 0.8 micrometres in diameter and yet is possibly the most abundant and successful organism on Earth. Just a single millilitre of your average seawater contains over 100,000 individual bacterial cells, and across the planet there are estimated to be at least an octillion individuals (10^{27}). We can safely say this is a very successful organism and its success is down to its ability to use sunshine and two of the most abundant molecules on Earth – carbon dioxide and water – to create sugar for it to live off.

Above: This engraving shows a study of the respiration of plants, wherein oxygen released by the plant displaces water under a bell jar.

Right: Though heliobacteria are phototrophic, meaning they convert light into energy, they can also grow without light by fermentation of pyruvate. They are exclusively anaerobic, and produce nitrogen, which they fix in soil and water.

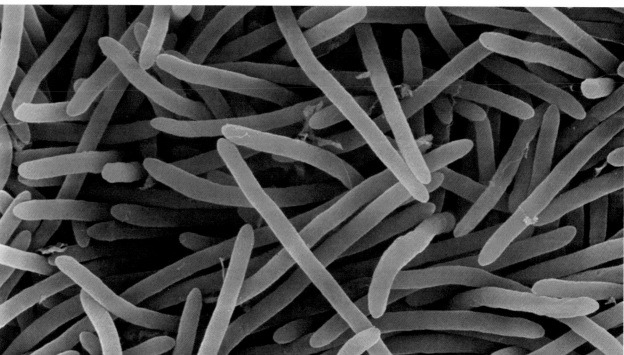

Right: The stomata of a lilac leaf (*Syringa vulgaris*). Stomata are pores that open and close in order to regulate gas exchange in a plant.

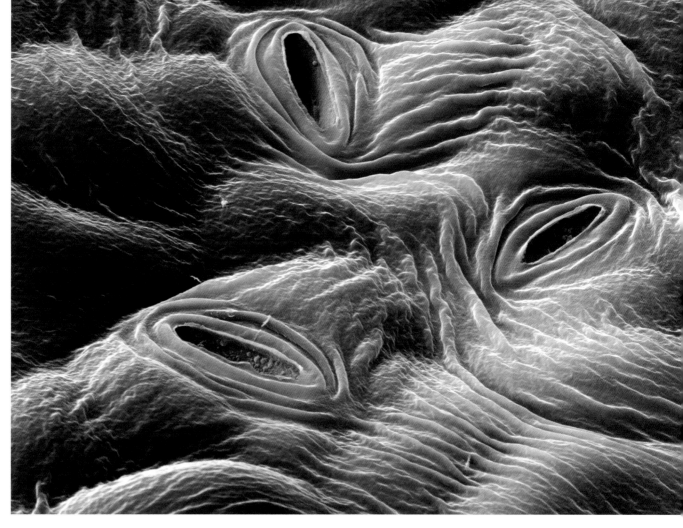

Below: *Monotropastrum humile*, seen here in Amami, Kagoshima, Japan, is a unique plant that lacks chlorophyll and is therefore unable to perform photosynthesis as most plants do; instead it gains sugars and nutrients from mycorrhizal fungi.

As far as we know, this ability to use water as an ingredient in photosynthesis (as the source of electrons driving the process), and hence produce molecular oxygen as a waste product, only evolved once in the history of this planet – and it did so as a single direct descendant of the cyanobacteria that we still find living on Earth. These are part of an unbroken chain of life that leads back to a time around two and a half billion years ago when somehow the biochemical machinery to perform this chemical magic trick came into existence. We do not know exactly when this game-changing form of photosynthesis first emerged, as the fossil record of cyanobacteria only really kicks in about two billion years ago, but whenever this trait did evolve in one of those tiny bacteria, it set in motion a series of events that would change the planet forever. Today, cyanobacteria are the only prokaryote (simple single-celled organism) that can perform oxygenic photosynthesis, and yet every plant on the planet owes its ability to photosynthesise to those first bacteria.

Above: Stromatolites are mineralised microbial communities, formed from cyanobacteria over thousands of years as the cyanobacteria trap detritus and sediment to create large living rafts known as microbial mats. The cyanobacteria also secrete calcium carbonate, which causes the mats to mineralise and form rock-like structures.

Right: Living stromatolites like these, photographed in Western Australia, are found in only a few salty lagoons or bays on Earth.

'There are these moments in the history of life that seem to have only happened once. Oxygen-producing photosynthesis is one of them. Was it a freak accident? We just don't know. Suddenly the oceans became the fuel, and this allowed life to scale up at least ten-fold.'
Professor Nick Lane, University College London

It's amazing to think that every single plant and algal cell on the planet has acquired this ability directly from those first photosynthesising bacteria over two billion years ago. The reason for this is not because those bacteria evolved directly into plants; instead, we think that in a moment of biological serendipity a cyanobacterium entered a primordial eukaryotic cell (the cells that all complex life are built from) in a process known as endocytosis, carrying its photosynthetic ability into the heart of the cell. In doing this it created the first plant cells on Earth. We now call these 'ingested' cyanobacteria chloroplasts, which are the site and organ of photosynthetic activity in all plant and algal cells. In every sense of the word we really do live on a planet; in fact, we are the direct product of a planet that has been shaped by the tiniest of living things into the endless complexity of life of which we are part.

RED EARTH

Before any of this complexity could burst across the planet, the Earth itself and the atmosphere clinging to it had to be transformed. Our great protector, the rich and complex balance of gases above our heads, had to be created and sustained. And yet two and a half billion years ago, when the first photosynthesising cyanobacteria began proliferating across the planet, the amount of oxygen in Earth's atmosphere was less than 1 per cent. This was still in human terms a toxic planet, with an atmosphere made up of predominantly nitrogen and carbon dioxide.

Slowly, as new Sun-loving life forms began to thrive, that single change in that single cell allowed those first photosynthesising bacteria to multiply at an astonishing rate, outcompeting almost everything else. Within a few million years, trillions of these photosynthesising bacteria were spread across the oceans, reproducing, photosynthesising and generating substantial amounts of oxygen gas. At first glance this would appear to be the point at which a new oxygen-rich atmosphere would have begun to emerge on Earth, but this was not an atmosphere that would transform in an instant. This was because despite an endless stream of oxygen being released into the oceans from the ever-increasing colonies of bacteria, the tiny bubbles would dissolve and vanish before they had a chance to reach the surface, captured by the water in which they are born.

In the Archean Eon life was prospering, but Earth seemed destined to stay a toxic world shrouded in a choking orange atmosphere. It was a living world but essentially one that was in stasis. Life remained microscopic, limited to the oceans with an atmosphere that was still almost entirely devoid of oxygen, but around two billion years ago this balance was about to shift again. As the cyanobacteria flourished, the levels of dissolved oxygen in the oceans slowly crept upwards; trapped, but far from docile, this new element didn't just sit there passively, it set about doing exactly what oxygen always does: it began to react.

Right: *The Angel of the North* turns red with exposure to the air, as the steel (an iron alloy) from which it is made oxidises into rust.

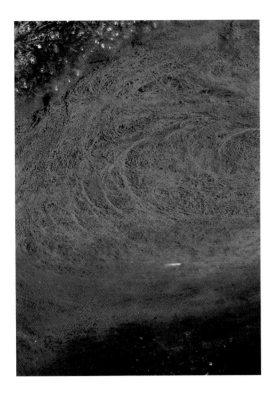

Above: Iron oxide coagulates and fluctuates at the surface of water in a small rivulet. Earth's early oceans were once similarly filled with iron.

Right: The Statue of Liberty is made from copper, which turns green when it oxidises rather than red, as iron does.

'The evolution of our atmosphere is in many respects the story of the evolution of life on our planet. Life can change a planet fundamentally. It is always this cause-and-effect kind of dance between the environment changing life and life changing the environment.'
Professor Tim Lyons, University of California, Riverside

Just like today, two billion years ago Earth's oceans were packed with dissolved minerals, but unlike today there was relatively little oxygen to react with them. That meant Earth's water was also packed full of dissolved iron and other metals.

Iron, like oxygen, is invisible to us when it's dissolved in water, but we all know what happens when iron, oxygen and water come together – we get rust. It's a process that we see occurring around us all the time. Rusting is the seemingly simple reaction of iron and oxygen in the presence of water to form iron oxides and hydroxides. It's a reaction whereby the iron acts as the reducing agent (gives up electrons) while the oxygen is the oxidizing agent (gains electrons) to form the new compounds.

Other metals besides iron undergo similar corrosion, but it is only the oxidation of iron that is known as rust. The result of this everyday chemistry is the reddish-brown, flaky material that we see pervading so many of the metal objects we leave out in the open. We see the effects of this universal chemistry almost everywhere in the modern world and we have to endlessly guard against the weakening effect it has on the structure and function of big and small items, from a bicycle chain to a bridge. The reason it is so damaging is that the brittle, flaky iron oxide takes up more volume than the original metal, generating forces that have caused structures to collapse and brakes to fail. And it is water that is almost always driving the reaction, because it is the catalyst that seeps into the microscopic cracks that occur on any iron surface, where it forms tiny puddles of acid that expose more and more metal to the destructive effect of the oxygen in the air. Seawater, with its abundance of chloride ions, speeds up this effect even further, creating the corrosion that is so prevalent in coastal and maritime structures.

'Copper minerals, blue and green minerals, yellow and orange minerals, uranium minerals, molybdenum, minerals of nickel, minerals of cobalt, all the colours of the rainbow represented in minerals formed because of the Great Oxidation Event.'
Dr Robert Hazen, Carnegie Institution for Science

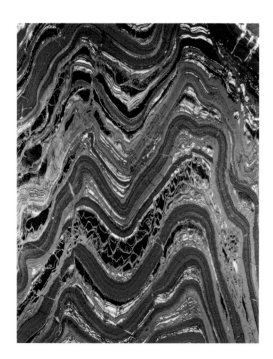

Such ubiquitous chemistry appears to be part of the very essence of our planet's makeup today, but this hasn't always been the case. Without free oxygen there can be no reaction and therefore no rust, and that's exactly what the Earth was like for the first two billion years of its existence. Now, I'm not suggesting you should suddenly be celebrating the introduction of rust into our planet's chemistry set for no reason, because as mundane as it sounds, the arrival of rust was transformative in the life story of the Earth, during an event that has the catchy name of the Great Oxidation Event, the Oxygen Catastrophe or the Oxygen Crisis (take your pick).

Driven by the tiny photosynthesising cyanobacteria and their insatiable appetite for sunlight, two billion years ago the Earth's oceans had reached a tipping point. The levels of dissolved oxygen had risen and as a result under the surface of the water something extraordinary was happening – it was starting to rain rust. As the oxygen and iron interacted, rust poured down onto the sea floor, turning the world's oceans into an eerie shade of red.

We see the evidence of this grand event in some of the most beautiful and striking rock formations on Earth. Banded iron formations were first discovered by prospectors looking for sources of iron to mine in Michigan's northern peninsula, in the United States, back in 1844. Found in sites across the world from Australia to India, South Africa to South America, these distinctive sedimentary rock formations are made up of alternating thin layers of dark rock rich in iron oxides followed by lighter, redder bands comprising of iron-poor chert. The scale of these formations can be vast, covering areas extending across hundreds of kilometres and reaching thicknesses hundreds of metres deep. To look at one of these striated structures is to look at the very moment in time when all of that rust started to rain down into the oceans. Each layer is a product of the oxygen produced by the cyanobacteria in those early oceans, laid down over millions of years on the seabed, and as we look at these layers we are seeing how the success of that early life waxed and waned, changing the composition year after year. As we can see from the existence of these vast rock formations across the globe, this process of oxidation occurred on a massive scale, but ultimately there was a limit to the amount of iron dissolved in the oceans and there was only so much oxygen that could be locked away. Eventually the vast chemical reaction playing out in the oceans would run low on ingredients, and with that decline the course of the Earth's history would change once again.

For half a billion years, oxygen, the photosynthetic waste product, had been trapped in the oceans. But now, with the iron stores turned to rust and floating to the bottom of the seas, for the first time in Earth's history oxygen had the chance to break free. With the oceans already reddened with this rust, it was the turn of the land to come under attack. Over just a few million years oxygen flooded into the atmosphere, immediately leaving its tell-tale reactive signature directly on the planet as the levels of free oxygen climbed.

Left: The folded layers of a banded iron formation were laid down in shallow seas where primitive bacteria caused iron to be oxidised and precipitated. Subsequent deep burial of the rocks, and tectonic movements, have caused the rock to alter and deform. This small sample is around 3 billion years old.

Opposite: The Rainbow Mountains in Gansu Province, China, are beautiful examples of banded iron formations.

The Great Oxidation Event was fully underway and it would change the geology of our planet forever. This is because the released oxygen literally began to eat the Earth. Rocks with iron and aluminium contained within them begin to rust and crumble into dust, driving vast dust storms that whipped across the planet. Our young world was being ripped apart and reshaped by its own atmosphere, and just as we find the evidence of this Great Oxidation Event in the ancient seabeds of the banded iron formations, so we find the evidence of the land's explosion of chemistry in rock formations all over the world, providing us with direct evidence of the action of all of those trillions of cyanobacteria churning out oxygen on the land as well as into the sea.

Before the oxygen arrived, the planet was monotone, barren and painted in shades of grey and black, but as the oxygen levels increased in the atmosphere this all changed. Oxygen and carbon dioxide, combined with water in the air, created acid that ate away at the planet's surface. It triggered an explosion in the formation of new mineral compounds, creating an estimated 2,500 of the 4,000 minerals found on the planet's surface today: many of them forming for the first time not just on our planet but in the Solar System. The result was a landscape transformed into a technicolour marvel.

Left: Iron, aluminium and potassium sulphates crystallise to form almost botanical shapes. Though plants were still a way off, the oxygenated Earth began to look more organic and much more colourful.

Below: An illustrated idea of how the Earth may have appeared in the Archean Eon (4 to 2.5 billion years ago). At this time the seas would have been filled with stromatolites, which were pumping oxygen into the atmosphere and would alter the planet forever.

Opposite: A large field of manganese nodules on the floor of the north-east Atlantic Ocean. Manganese nodules form when manganese and iron minerals are deposited on the surface of an object, such as a tiny piece of bone. The minerals in the seawater deposit over time; a nodule grows by 1 centimetre diameter in a million years.

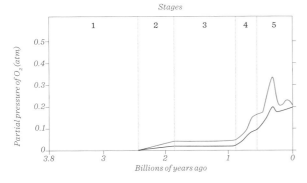

GREAT OXIDATION EVENT

O_2 build-up in the Earth's atmosphere. The red line represents the highest estimated figure for the partial pressure of O_2 and the blue line the lowest. The PO_2 today is 0.210 atm. Time is measured in billions of years ago.

Stage 1: 3.85–2.45 billion years ago
Practically no O_2 in the atmosphere. The oceans were also largely anoxic.

Stage 2: 2.45–1.85 billion years ago
O_2 produced, rising to values of 0.02 and 0.04 atm, but absorbed in oceans and seabed rock.

Stage 3: 1.85–0.85 billion years ago
O_2 starts to blast out of the oceans, but is absorbed by land surfaces. No significant change in oxygen levels.

Stages 4 & 5: 0.85 billion years ago–present
Other O_2 reservoirs filled; gas accumulates in atmosphere.

Today, the effects of this burst of oxygen are painted everywhere on Earth, but it wasn't just the impact on the land that arrived with this atmospheric revolution. The climate changed dramatically with the arrival of oxygen. Atmospheric methane that had clung to the planet for millions of years, blanketing it with its strong greenhouse characteristics, became oxidised into carbon dioxide, a weaker greenhouse gas that caused the planet to begin to cool. But perhaps the most profound impact of the oxygen was its effect directly on life. The sudden injection of oxygen into the atmosphere was less a breath of fresh air and more the kiss of death for much of the life that had evolved in the anaerobic (oxygen-free) biosphere that had existed for billions of years. This oxygen blast triggered a mass extinction event that cleansed the planet of many of the anaerobic species that had flourished up until now. But as always with life, as one road closes another opens, and in this case the opportunity that emerged with the arrival of oxygen was truly transformative. As the oxygen levels increased, the chemistry of the oceans and the land transformed, and so too did the chemistry of life. Held back for billions of years by the energetic limitations of anaerobic respiration, the arrival of oxygen, with its unique chemical characteristics, allowed the handbrake to be taken off. Now, with access to vast new reserves of energy, life was able to flourish in ways that had been unimaginable in the previous three billion years. From destruction complexity emerged, because as the oxygen levels in the atmosphere climbed, the opportunity transformed for a life form that had until now been hidden in the shadows.

'The very composition of the atmosphere itself was changing. This was known as the Great Oxidation Event. All of a sudden, new chemical possibilities arose all over the surface of the Earth, and amazing changes were to follow.'
Dr Robert Hazen, Carnegie Institution for Science

Aerobic life, oxygen-loving, complex cells know as eukaryotes, are thought to have first emerged around two billion years ago. A form of life entirely different to the domains of bacteria and archaea, this is life that relies on oxygen to power its metabolism. With a nucleus full of DNA, these cells also had the ability not just to exist as single-celled organisms but to combine to produce organisms of increasing complexity; multicellular organisms that can form into different specialised cells that form the tissues, organs and systems that underpin the complex biology of our planet. The impact on life would not be immediate, but as we will see in the next chapter, with the arrival of oxygen into the atmosphere the age of animals, plants and fungi had begun.

Opposite: The Maunsell Forts are armed towers built in the Thames and Mersey estuaries during WWII to help defend the United Kingdom. Built from iron, they are beginning to rust as the iron oxidises.

Above: This mesmerising mineral landscape of orange and red fairy chimneys, or hoodoos, in Bryce Canyon National Park, Utah, USA, indicates high levels of iron and other minerals in the rock.

INTO THE BLUE...

There was one more thing: oxygen transformed our oceans, reshaped our land and became a vital ingredient for complex life today. But this volatile gas was also to have one other beautiful and dramatic impact on our planet. As oxygen enriched the atmosphere and stripped away the methane that had hung for so long in the air, it meant that after two billion years of fog the thick orange haze was lifting. For the first time in Earth's history the skies began to clear to reveal a new phenomenon – a sky where oxygen and nitrogen scattered the white sunlight hitting the Earth's atmosphere to create a brilliant, breathtaking blue. It was the final calling card of our planet to the Universe, to signal that this was a living, breathing planet, a planet where organisms capable of complexity not only existed but thrived, a planet that was rich in water and in life. All of this was held in place by the great protector, the thin blue line of gas that marked the place where the hostility of space could be held at bay to allow this floating ball of rock to become a home.

Life had become a powerful force, able to alter the course of the planet's climate and geology. And 1.2 billion years ago, more complex life was still evolving. But this placid green and blue Earth was just about to be thrown into the deep freeze, as the whole planet turned white. On snowball Earth, the future of life would hang in the balance.

'We have immense power over our planet. Our atmosphere is precious. And if we change it too much, the whole world will pay the price.'
Usha Lingappa, University of California, Berkeley

Right: After the Great Oxidation Event and the formation of the atmosphere, the Earth had its first blue sky. Now with early oceans as well, the once-barren rock had become a blue planet.

Opposite: A Martian sunset, photographed by the Spirit rover, manages a small blueish glow around and above the Sun, but without a strong atmosphere, and wreathed in thick dust, the skies above Mars appear reddish most of the time.

SNOWBALL

'To understand how life survived in ice
in the past, we need to look to how life
survives in ice today.'
Dr Jaz Millar, Bristol University

A WORLD OF CHANGE

We live in a world where almost daily we are reminded of the critical importance of ice. The melting of great glaciers and the disintegration of the polar ice sheets, as they fracture into the rising ocean, are powerful images illustrating the impact of increasing temperatures on our planet caused by human-induced climate change. We instinctively know that ice is one of the great barometers of our planet's health, and we know that the disintegration of our frozen landscapes is more than just a canary in the coalmine. Earth is transforming before our eyes, transmuting into a world that will be nowhere near as welcoming as the one to which we have become accustomed.

Above: During Earth's Snowball Event, the temperature at the equator was similar to that of modern-day Antarctica.

Right: The Athabasca glacier in Alberta, Canada, has lost 60 per cent of its ice in the last 150 years due to the rising temperatures.

Far right: Ice on land and sea is now melting at a rapid rate, as seen here in Greenland.

The balance of ice on our planet is a precarious thing, intimately linked to the range of temperatures that exist across the Earth. From the hot and humid equator, with an average afternoon temperature of 31 degrees Celsius, to the frozen poles, where in the Antarctic temperatures average at minus 60 degrees Celsius during the winter, only rising to minus 28.2 degrees Celsius in the summer. At first glance this appears to be a vast range, with an average temperature recorded across the planet of 13.9 degrees Celsius in 2020 at any one time, while the temperatures at the hottest and coldest places on the planet are 100 degrees Celsius apart.

But this seemingly large range is deceptive; for us it might seem like a broad shift from hot to cold, but in the context of a universe where temperatures can range from -273.15 degrees Celsius in the Boomerang Nebula (the coldest place in the Universe) to 300 million degrees Celsius in the RXJ1347 cluster of galaxies, our little ball of rock is kept in a tightly held temperature bubble. And as we have seen in the previous chapter, this is in large part due to the protective layer of air, the Earth's atmosphere, which keeps the harshness of space at bay by holding the warmth of the Sun like a blanket around our planet. Not too hot nor too cold, it's this precise temperature level that creates the balance between the liquid water that exists across the surface of about 71 per cent of Earth alongside the permanent ice caps that top and tail the planet.

It's only when we realise that we live in such tight temperature parameters in a universe of extremes that it becomes easier to understand why the smallest changes in temperature can have such profound effects on our planet. As we are seeing to such great cost, an average increase of just 1 degree Celsius can profoundly unbalance our world – ice melts, oceans rise, the weather intensifies and our precious biosphere contracts. We are, and always have been, at the mercy of the Earth's climate – and that's not just human life but all life forms. If you look back across the history of our planet there are endless examples of how life has been shaped by the changing climate.

The evidence for this is clear today, where just the smallest of shifts can cause Earth's climate to spiral into extremes. In this chapter we are going to tell the story of one of the most extreme examples of climate change in our planet's 4.5-billion-year history: a moment in time when the great forces conspired not just to shift the environment on Earth but to transform it from a living, breathing water world into a frozen wasteland, on the brink of being devoid of life.

Seven hundred million years ago these planetary forces were unleashed on an unimaginable scale, pushing Earth to the brink. From pole to pole, across all the lands and oceans, our entire planet became frozen in a climate disaster that couldn't have happened at a worse time.

The advancing ice threatened to destroy something incredibly precious, because as this big freeze rolled across our world's surface it encased the newly emerging, complex life forms, built from a new type of complex cell that we now know is the ancestor of all animal life, a branch without which we would not exist.

But this moment in Earth's history is not just the story of a near miss, it is also a tale of resilience, because this climatic and environmental catastrophe that threatened our living world with extinction ultimately led to something miraculous. You see, this new life didn't just survive the ice and its aftermath – it *thrived*. The deep freeze saw a new age begin on Earth, one of complex life and of a greater size and diversity than had ever been seen before. Life that would go on to dominate the oceans and the land to this very day.

LOST WORLDS

RODINIA SUPERCONTINENT
This reconstruction depicts the supercontinent Rodinia 750 million years ago. The blue shows orogenic belts – areas where tectonic plates are pushing against each other.

We are so used to the map of the world that has been imprinted in our minds that it's very difficult to imagine it looking any other way. But this atlas that appears so timeless to us is in fact incredibly fleeting. Just over a billion years ago, the face of our planet was completely different. Whereas today we have five continents spread across the globe, back then all of the Earth's landmasses were gathered together into one giant continent we have named Rodinia.

One of the earliest supercontinents, this vast landmass dominated the planet from its formation approximately 1.2 billion years ago and over the next almost half a billion years. Moulded from the collision of leftover fragments of other supercontinents that had split apart, Rodinia was covered in an endless array of geological beauty. From jagged mountain ranges to vast floodplains, it's easy to create an image of this lost world in our imaginations, until you realise this was a barren planet, devoid of any plant or animal life. With little oxygen in the atmosphere and no ozone layer to protect any life near the surface, Rodinia was sterile, existing at a time before complex life had been able to gain any grip on dry land.

Today, the large, complex life that we see all around us can only really exist if it's part of a bigger, interconnected ecosystem, one that is as complex as the life it supports. But the truly staggering thing is that life like this was nowhere to be found for over 4 billion years – that's pretty much our entire planet's existence. The only life forms that existed 800 million years ago were found around Rodinia's coastal waters. Here, vast bacterial mats stained the ocean floor, while blooms of cyanobacteria flourished on the periphery of the supercontinent, all built from prokaryotic cells – the type that has no nucleus or ability to form anything other than a single-cell organism. Despite life having existed on the planet for over 3 billion years, this seemed to be as complex as it got – colonies of bacteria being the best this living planet had to offer.

But lurking among these bacteria, perhaps lost in the crowd of simple microbial cells, was a new form of life, one that would become the basis of the endless complexity of life forms we see on our planet today. A new type of complex cell, the

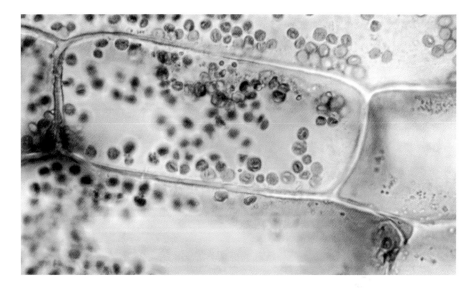

Right: This magnified image of plant cells shows chloroplasts bounded by the cell walls.

Right below: This cross section of an olive leaf shows how the structure is designed for the transport and retention of water. The large central bundle of tubes is where water is circulated. Below the upper epidermis and surrounding the vascular bundle are the cells containing chloroplasts. Beneath this is the spongy mesophyll, with large intracellular spaces for gaseous exchange. At the bottom is the lower epidermis, a single layer of closely packed cells where the stomata control gas exchange.

Prokaryotic Cell

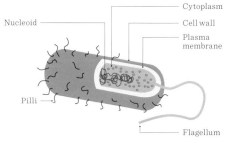

Eukaryotic Cell

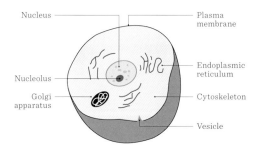

PROKARYOTES VERSUS EUKARYOTES

All life on Earth can be sorted into two groups depending on the fundamental structure of their cells: prokaryotes and eukaryotes. Prokaryotes have cells that lack a nucleus or any membrane-encased organelles, while eukaryotes have cells with a membrane-bound nucleus that stores genetic material as well as membrane-bound organelles. Prokaryotes on Earth include bacteria and archaea. Eukaryotes include animals, plants, fungi and protists.

Opposite: Around 800 million years ago, the coastal waters around Rodinia would have looked much like this image of a cyanobacterial bloom in the Netherlands.

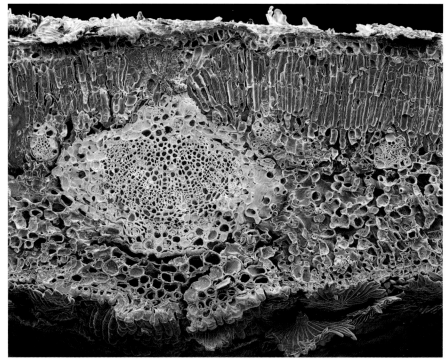

eukaryotic cell (from the Greek *eu*, meaning good, and *karuon*, meaning nut), had the characteristics that would allow it to exist not just as a single cell but as part of complex multicellular organisms. With these eukaryotic cells life could develop specialised cells to create structures and organs, and the internal systems that would in turn create a complexity of life that could form the interconnected systems of plants and animals, predators and prey into an ecosystem of complex life that we see all around us today.

This abundance of life would come from the evolution of this one type of cell, but the origins of this biological revolution remain, like many of the biggest moments in the history of our planet, resolutely hidden from view, because we still don't know when these first complex cells evolved on Earth. But what we do know is that all animals, plants and fungi evolved from this same basic eukaryotic cell type.

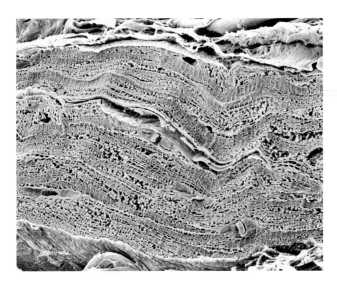

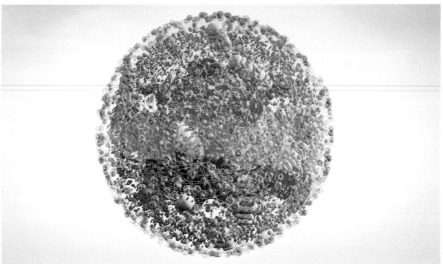

Above: The mitochondria inside a cell are where chemical energy is produced and stored as a small molecule called adenosine triphosphate (ATP).

Right: A human cell contains all the key components of a eukaryotic cell: nucleus, organelles, mitochondria and more – it's a complex powerhouse.

'Although the supercontinent break-up could have helped fuel eukaryotic life, it also set into motion a series of events that could have led to catastrophe for them.'
Dr Michael Wong, Carnegie Institution for Science

Today, we see these cells everywhere; we are made up of them, just like every other life form that you can see. The feature that defines these extraordinary cells and differentiates them from their prokaryotic cousins is the presence in the middle of the cell of a nucleus, the structure that contains within it all the genetic information, the genome of that life form. It's here that the instructions to transform a single cell into a multitude of specialised cells and structures resides – but that's not the end of the eukaryotes' complexity. Found around the nucleus are other structures unique to eukaryotes that power the cells' metabolism, such as mitochondria and, in the case of plants, chloroplasts – the powerhouses behind photosynthesis. It's this cellular machinery that makes eukaryotic cells so much more powerful and versatile than the simpler prokaryotic cells of the bacterial kingdom, enabling them to build a seemingly endless variety of multicellular organisms that can form rich and complex ecosystems.

We have discovered that prokaryotic cells can be dated back to the very beginning of life on Earth, emerging at least 3.8 billion years ago, but the origin of the eukaryotic cell is a moment in the history of life that remains harder to pin down. Some evidence has pointed to the emergence of eukaryotic life as far back as 2 billion years ago, but this is far from substantiated and many scientists believe the first appearance of these specialised cells was probably much later than that. More reliable fossil evidence seems to suggest that eukaryotic life in the form of a filamented red alga was living in coastal waters at least 1.2 billion years ago, just as the supercontinent Rodinia was beginning to form. Whether this alga was among some of the first complex life to exist or not, we do not know, but eukaryotic life at this time was still far from dominant. These pioneering life forms from which so much complexity would ultimately emerge were struggling to scrape out an existence. This was still a hostile environment, with no complex food web to sustain it, so although this early eukaryotic life was able to survive, in order to thrive these energy-hungry life forms would need far more food and nutrients than this simple ecosystem could provide.

Stuck on the margins of the land, the ancient ancestors of all complex life were locked in a world with no obvious way forward. This is how it might have stayed; without any significant change there was no hope of the world we see today ever appearing, but, luckily for us, the status quo that Rodinia had helped maintain for almost half a billion years was about to be shattered.

Right: The Tirohan Dolomite rocks in central India contain fossilised microbes from ancient stromatolites. These two samples show *Ramathallus lobatus*, a multicellular red algae dating from 1.6 billion years ago.

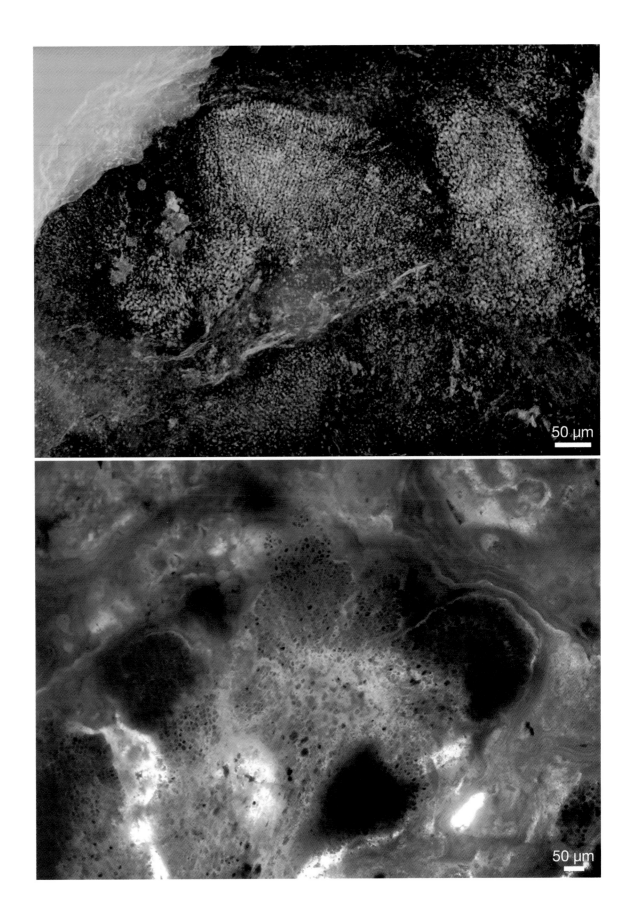

Walking through places like this, you get a real sense of the raw, primal, brutal forces at play – and they are at play. You see, the walls of this chasm are moving apart by one centimetre each year. Now it doesn't sound like much, but that's because to understand the way that these forces impact on the landscape, we've got to think about it in geological time. That's hundreds of millions of years, not hundreds of years. The rock beneath is literally being torn in two along a fracture that runs for thousands of kilometres.

EARTH SHATTERING

TECTONIC PLATES

About 40 kilometres northeast of the Icelandic capital Reykjavik, in the Thingvellir National Park, lies one of the true wonders of the natural world. It's almost impossible to imagine a documentary series on the history of the Earth not visiting this geological marvel, and it was no surprise when we turned up with our camera crew to find we were not alone, as we joined a stream of pilgrims coming to pay their respects at this geological shrine.

Lying in a rift valley that marks the boundary between two giant chunks of the Earth's crust, Thingvellir is littered with the evidence of the Earth having literally ripped itself apart. This is one of the best (and one of the few) places on the planet where you can see two continental plates – in this case the North American and Eurasian tectonic plates – pulling themselves apart without having to dive miles down to the ocean floor in a submersible. This place is littered with the evidence of tectonic plate movement, which has created the beautiful cracks and water-filled canyons that draw so many to the area. To stand here along this divide and witness the force of these two plates being dragged apart is quite frankly mindblowing; you can literally see (and sometimes feel from the not-infrequent earthquakes) how the land is being stretched and torn apart.

This phenomenon is seen not just in Iceland but also in Africa, where the same process is creating the great East African Rift Valley. This split stretches for over 3,000 kilometres, all the way from East Africa through to the Red Sea, and is widening by around 2.5cm each year. Standing in a location like this is a visceral reminder of the fact that whether it's right in front of our eyes or far below in our oceans, the surface of our planet is continually being torn apart by the forces churning deep within it. And these forces have been moving the world's greatest landmasses, creating and destroying them, for more than 3 billion years – an irresistible power that nothing, not even a giant supercontinent, can resist.

Over millions of years the same forces that are today tearing our world apart wrought their slow-motion destruction across the supercontinent of Rodinia. Great fault lines cleaved open this ancient Earth, signalling that after dominating the surface of the planet for almost half a billion years, Rodinia's days were finally numbered. The boring billion, as the preceding billion years have become known, were coming to an end, and this monolithic continent that had endured for 350 million years was slowly but surely being ripped apart. This was a continental change on a scale that the world rarely sees, during which the face of the planet was being reshaped into something entirely new. But contained within this destructive power was also opportunity, created by something with the power not only to transform the land but also life, triggering the transition from that barren planet to our modern world.

In many ways, we can think of the history of our Earth as being about the interplay between life and geology. Sometimes the great planetary processes have given life a bit of a leg up on others that they've knocked right back down again. What's interesting is that the moments when the trajectory has changed for both life and the planet together are few and far between. One of these moments was about to play out, and what happened next would go on to define the future of our entire world...

Left: A diver just above the tectonic boundary between the Eurasian and the North American plates, at Thingvellir National Park, Iceland.

Opposite: The Almannagjá gorge, a surface crack of the mid-Atlantic Rift in Thingvellir National Park, Iceland.

Above: This earthquake fault
line in Arizona, USA, shows
how the continental surface
can shift.

As the epic forces slowly dragged the continent apart, it created deep valleys
and towering mountain chains that not only transformed the landscape but the
ecological balance, too. Geological metamorphosis on this scale supercharges the
processes of erosion, causing a flood of minerals and nutrients to pour off the land
and into the oceans, making the world of our distant ancestors suddenly a much
more nurturing place to live. As this rich stream of new resources flooded into the
oceans, life exploded, radically increasing in number and variety, and the whole
ecosystem burst forward with a new and complex diversity, all driven by the endless
adaptability of the eukaryotic cell.

This marked the beginning of an intricate web of life, the first echo of those
around us today. Kick-started by the changing environmental conditions, we think
this explosion of life was then propelled forwards by the start of one of the most
powerful drivers of evolution. In this new dynamic environment we think the first
predator–prey relationships began to evolve. It triggered a race for survival that
caused an explosion in form and complexity that has continued in an unbroken
chain to this day. But don't think for a moment that this was the grand predator–
prey relationships we see playing out across the planet today, because this cat-and-
mouse game was hidden in a microscopic world.

This page: Predators and their prey are key to the web of life on Earth. Clockwise from top left: cheetah hunting down gazelle; osprey with fish; bear and salmon; jumping spider and cockroach.

'The fossil structures made by early eukaryotes are microscopic, but we can see spikes and spines. It's energetically costly to make such structures; we think the organisms made these as defensive weapons to protect themselves from being eaten.'
Professor Phoebe Cohen, Williams College

The evidence for this fundamental shift in animal behaviour that coincided with the breakup of Rodinia has come from tiny fossils discovered in rock formations that sit in a part of the Grand Canyon in Arizona known as the Chuar Group. This layer of rock was once an ancient seabed that existed in the Neoproterozoic Era between 782 and 742 million years ago, just as that vast supercontinent was being dragged apart. This is, in effect, a directly preserved record of those nutrient-rich oceans that burst into bloom around the edges of Rodinia as it began to die.

A group of scientists led by palaeontologist and geobiologist Susannah Porter, from the University of California Santa Barbara, used a scanning electron microscope to examine minute fossils of single-celled amoeba life forms they extracted from this ancient rock layer. Although these life forms are smaller than a grain of sand (measuring 75 to 150 micrometres), the microscopic investigation revealed perfectly circular holes that appeared to have been drilled into these single-celled creatures. Porter and her team believe these are scars left on the individual amoeba cells that were, it seems, unlucky enough to have become someone else's lunch. The suggestion is that these flawlessly drilled holes may have been made by some of the first predatory creatures living in those warm, nutrient-rich, shallow seas some 740 million years ago. This is the oldest evidence of predatory behaviour in eukaryotes ever discovered, and at the time of writing it is the earliest evidence of predation in the fossil record.

The circular holes in a microfossil from the Chuar Group are thought to have been formed by predatory, vampire-like protists that drilled into the walls of their prey. These predatory holes uncovered in these investigations are strikingly similar to the wounds left by a modern-day predatory amoeba, with the delightfully dark name of *Vampyrellid amoebae*. In these contemporary species the amoeba uses a special appendage to lock onto part of the prey's outer surface, and it then produces an enzyme to cut a ring through the cell wall, providing direct access to the nutritious contents of the cytoplasm within. We can't be certain that the tiny fossils discovered in those ancient rocks of the Grand Canyon were early adopters of this kind of predatory technique, but they do provide the first direct evidence to suggest predation was one of the driving factors in the diversification of eukaryotes that took place about 800 million years ago. According to Professor Susannah Porter, 'This is giving us a first glimpse of the diversification of complex cells – eukaryotes – rather than bacteria and archaea [a group of single-celled organisms with no nucleus]. It shows us that our ancestors had started becoming important in terms of their role in ecosystems and their diversity on Earth.'

Collecting all the evidence together, it appears that a new era of vitality and diversity had blossomed in the shadow of Rodinia's disintegration, but as is almost always the case with the history of life on Earth, the stage that had been set was far from stable. For all of the impetus that Rodinia provided to life, in its death throes it was also about to deliver a near-fatal blow to the trajectory of complex life forms. Just as life was taking this great leap forwards, the forces tearing the ancient supercontinent apart were about to turn against it, plunging the entire world into a seemingly endless, desolate winter.

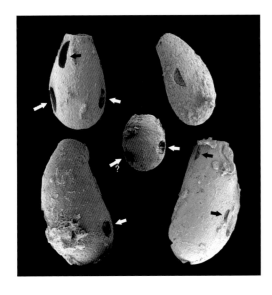

Opposite: Chuar Butte in the Grand Canyon, Arizona, USA, is composed of Permian Kaibab limestone, which overlays cream-coloured, cliff-forming, Permian Coconino sandstone. These formations were caused by uplift along the Butte Fault, which sits along their western edge.

Above: This microscopic image from Susannah Porter's investigations in the Chuar Group region of the Grand Canyon shows the fossils of single-celled amoeba that have been drilled into by some of the first predatory protists living in a warm, shallow sea some 740 million years ago.

FIRE AND ICE

Below: Lava flows create new basalt plains in their wake.

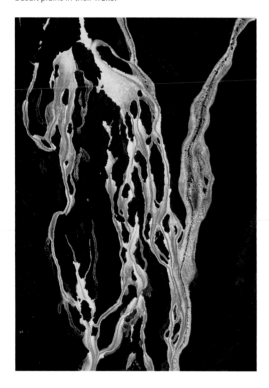

Above: Volcanic eruptions introduce a wealth of minerals into the air and soil around them, making volcanic regions ideal for growing many types of food. The volcanic soil of the Canary Islands in Spain has been optimal for wine production for more than 500 years – even Shakespeare references a 'cup of canary' in his play *Twelfth Night*.

Right: The hand-cultivated black volcanic soil on the slopes of Volcan Zunil above the village of Xetinamit, in Guatemala, is ideal for growing potatoes.

This was a winter that began not with ice but with fire. As the great continent of Rodinia continued to break apart, its destruction created a huge pulse of violent volcanic activity, and it was in the maelstrom of this fire that the seeds of the great freeze were first sown.

This was volcanic activity on an unimaginable scale, with eruptions lasting millennia, spewing out over 2 million square kilometres of lava that would transform the surface of the Earth. But it wasn't in the aesthetic of this rock that Earth's transformation would come – it was in its chemistry. As all this lava cooled, it created a vast black stain on the dark, jagged rock basalt across the heart of the fracturing continent – one that would have an astonishing effect on global temperatures.

In July 2020 a study was published in the journal *Nature* with the typically unassuming title (for a scientific paper) of 'Potential for large-scale CO_2 removal via enhanced rock weathering with croplands'. Based on research carried out by a team of scientists across the world, led by the University of Sheffield in the UK, the title hides an incredibly simple but fascinating suggestion for tackling the climate crisis by reducing the levels of carbon dioxide in the atmosphere. Put simply, the suggestion was that by spreading a particular type of rock dust over huge expanses of global farmland you could literally suck billions of tonnes of carbon dioxide from the air each year.

This is by no means a solution, or anywhere near as effective as actually reducing carbon emissions, but it definitely could offer a potential short-term quick-fix to help reduce the rapid rise in global temperatures. All you need is a ready supply of rock dust containing basalt, the igneous rock that is formed from the rapid cooling of low-viscosity lava. This is possible, as more than 90 per cent of all the volcanic rock on Earth is basalt, forming volcanic island chains such as Hawaii and covering vast swathes of the surface of the Earth from Siberia to South America. This rock dust rich in basalt also already exists in enormous stockpiles from many mining processes that generate it as a waste product.

The rock dust approach, or Enhanced Rock Weathering as the scientists called it, is based on a simple piece of chemistry. Compared to the far more resilient rocks like granite, basalt deteriorates relatively rapidly when exposed to the environment and it is this process of weathering that not only changes the look of any rock but also its chemistry. This is because the chemical effect of weathering results in the

EARTH'S ICE AGES

A simplified representation of the five major ice ages that have occurred in the past 2.4 billion years of Earth's history.

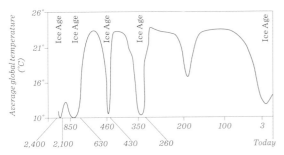

Below: While the last true ice age was nearly three million years ago, there have been mini ice ages since. One followed the decimation of the American population after its first contact with Europe (the subsequent significant increase in tree cover in once-inhabited areas led to increased oxygen levels and a global temperature drop). In the seventeenth century the River Thames in London regularly froze for months every year and Frost Fairs were held on its surface.

release of water-soluble ions like sodium, magnesium and, most importantly in this story, calcium. When released from the rock this calcium readily dissolves in water and reacts with carbon dioxide that has dissolved in the water from the atmosphere to form $CaCO_3$ – otherwise known as calcium carbonate. It is acting as a carbon dioxide trap that in effect turns the destructive carbon dioxide into stone that will eventually be washed into the sea and locked away in the seabed. It's a process that is both good for the climate and good for the soil, which is why it is now being widely explored as a potential way of meeting key global climate targets.

This is all very interesting, but what's the relevance of this to our story of Snowball Earth? Well, as we know, carbon dioxide is a very potent greenhouse gas that has the capacity to lock lots of heat into our atmosphere, and the more we pump into the air, the more we see our global temperatures rising. But what's fascinating is that the reverse is also true – if we take carbon dioxide out of the atmosphere, temperatures fall. And through this exact process of calcium carbonate formation, 720 million years ago the erosion of millions upon millions of tonnes of basalt spewed up when Rodinia broke apart and began sucking the warming blanket of carbon dioxide out of the atmosphere, which in turn flicked the Earth's thermostat from a comfortable stasis towards a rapid chill.

Within the space of a couple of million years, the plummeting levels of carbon dioxide had awakened a sinister force, hundreds of miles north of the riot of eruption in Rodinia. Something not seen on planet Earth for a billion years was appearing. Growing in magnitude in the Arctic Ocean, ice was returning, covering the poles of the planet and gradually making its way to more southerly latitudes. And as the temperatures continued to tumble and the planet slipped into a deadly chill, those distant ancestors of ours, the complex life that was just taking hold in the ancient oceans, was about to face its greatest challenge yet.

LIFE ON ICE

W hether you're microscopic or more human-sized, almost all life forms love some warmth. We all know the feeling of venturing out into the cold and sensing the threat to life that comes with even the shortest unprotected exposure to the elements. Humans are essentially creatures of the tropics, and with a constant need to maintain an average internal temperature of 37 degrees Celsius, we go to great lengths to protect this vital number – whether it's in our clothing, shelter or the heating that we generate to protect ourselves, it's an essential part of being an animal with a global footprint. When just a two-degree drop in core temperature can result in hypothermia and ultimately lead to death, it's no surprise that this maintenance of warmth is built into our behaviour and, of course, our physiology as well. When faced with the cold, our bodies shunt blood away from our skin through the process of vasoconstriction and we start to shiver, an action that generates heat in an attempt to raise our core body temperature.

But humans are far from alone when it comes to vulnerability to the cold. Across the living world, whenever temperatures drop sufficiently low for long enough, life slowly fades away. You just need to take a walk up to a few metres of altitude to see that life just cannot cling on as the temperature drops. There is no generally accepted value for the lower limit of temperature for life on Earth, but what we do know is that very few multicellular organisms can complete their lifecycle at temperatures below -2 degrees Celsius. Some bacteria and unicellular eukaryotic life forms have also been found to survive temperatures down to -18 degrees Celsius in hyper-salinated polar lakes that remain unfrozen due to the high levels of salt. But at these temperatures survival is the only success, and any kind of active cell division or growth has to wait for the summer months and the warming Sun. Plant life is equally restricted by the cold, with 17 degrees Celsius being the lowest temperature reported for photosynthesis to occur in plants like the Antarctic lichen, *Umbilicaria aprina*.

'Cold and ice can destroy cells and slow down chemical reactions, impacting life. It's likely that 717 million years ago, as the Snowball Earth glaciation began, microscopic webs of life were also threatened.'
Professor Susannah Porter,
University of California,
Santa Barbara

Opposite: Ice–albedo feedback is a positive feedback climate process. Ice coverage increases the planet's albedo so the Sun's warmth is reflected back into space rather than being absorbed, thus temperatures subsequently drop.

Right: Lichen are composite organisms made up of algae or cyanobacteria that are living among filaments of multiple fungi species in a mutualistic relationship. They were likely a very early life form on Earth.

Left: Large sheets of ice may have created the bottleneck required for the evolution of multicellular life on Earth during the Snowball Event.

Below: Today's temperatures only support ice sheets in the polar regions during winter, and even then, these sheets are becoming less solid as the climate changes.

'For us as humans Snowball Earth will always seem like a harsh, almost impossible planet to live on. However, when we look at the depths of Antarctica and the middle of the ice sheets in the Arctic, there is life wherever you look for it.'
Dr Jaz Millar, Bristol University

What applies to life today also applied to life 715 million years ago, as temperatures on the ancient Earth continued to fall. Across the fracturing continent great sheets of ice surged forth from the highlands of Rodinia, just as armadas of sea ice advanced from the north and south poles. The world of our ancient ancestors was rapidly shrinking as the ice tightened its grip on the planet. The idea that ice could swarm across the whole planet like this seems incomprehensible in a world where we are so used to its confinement, but we know it did because we've found evidence for the unstoppable advance of glaciers across the globe and the unimaginable destruction that comes with that. Ice leaves an indelible mark wherever it goes, and similar scars and debris to the ones we see around glaciers today have been found in 700-million-year-old rocks and landscapes all over the globe – evidence that suggests no part of the planet was able to escape the crushing advance of that ice.

The idea that the Earth might have gone through an absolute deep freeze has been a topic of conjecture for many decades. But as is often the case with the advancement of science, the evidence for this global event began to accumulate long before any theory existed. As far back as the 1870s, the first evidence of widespread ancient glaciation was uncovered in rocks in Scotland, Australia, India and Norway. It seemed that wherever geologists looked in the world they found evidence of massive glaciers that existed in the Earth's distant past, but without a full understanding of the endless movement of continents it wasn't possible to piece this evidence together into any kind of coherent theory.

It wasn't until the 1960s, when our understanding of continental drift and the journey of Earth's lost supercontinents began to develop, that Walter Harland, a British geologist, proposed the idea of a mass freezing of the Earth over 600 million years ago. Harland cited evidence that he and his team had uncovered in Svalbard, Norway, and also in Greenland, that suggested that samples of glacial till (a specific type of rock deposit that is generated and eventually deposited by the movement of a glacier) were originally deposited not in the far north but at tropical latitudes.

Left: Walter Brian Harland in Svalbard, Norway, in 1938.

Opposite: Sea ice was essential to the development of early life, and it's crucial to life forms like the polar bear today, but as it vanishes, these animals will have to adjust or die.

Above: The edge of a glacier coming off the Greenland ice sheet near Camp Victor. Glaciers are forces of destruction and change for landscapes.

Above right: Woolly mammoths were successful during the last ice age, but they died out as temperatures rose.

Below: A thin section of
oolitic limestone under the
microscope. It's made up
of calcified layers formed
around grains of shells or
quartz from ancient seabeds.

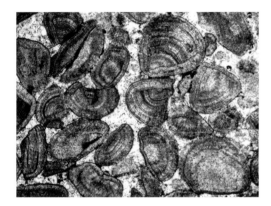

This was the first time that anyone had suggested an ice age so extreme that it had resulted in the frozen boundary reaching all the way to the equator.

Even then, geologists had a difficult time explaining how such an event was possible. What mechanism could have led to such extreme climate change? And what triggered it in the first place? As the evidence accrued, scientists became more and more convinced of the validity of the theory, but it wasn't until 1989 that Joe Kirschvink, Professor of Geobiology at Caltech, coined the term 'Snowball Earth' as a passing reference in a major study of the biology of the Proterozoic Eon. The name stuck, and although Kirschvink didn't continue this line of research himself, it did inspire other scientists, perhaps most notably Professor Paul F. Hoffman and his team, to run with the name and the theory. Hoffman gathered evidence from his fieldwork in Namibia that revealed ancient glaciation in rocks that were interwoven with limestone. Limestone only forms in warm tropical seas, so the discovery of glacial tilt and tell-tale dropstones sandwiched between layers of limestone suggested that for a period of time glaciers covered even the warmest parts of the planet during the Snowball Earth event. Publishing in the journal *Science* in 1998, here was the strongest evidence yet that the ice had no boundary, until eventually the warmth and ultimately the limestone returned.

Since Hoffman's work, the interest in the Snowball Earth theory has both intensified and become part of mainstream geological thinking, helped by the

'If you look underneath glaciers, you can see rocks trapped at the bottom of the ice. As a glacier advances out to sea, it carries these rocks with it, then as it melts, these big rocks at the bottom will drop into the deep ocean. Scientists call these dropstones, because once they hit the bottom of the ocean they become embedded in the seafloor and look different from surrounding layers. Over millions of years the ocean floor turns to rock and can be exposed as dry land.'
Professor Charlotte Spruzen,
McGill University

ALBEDO IN NATURE
The percentage of diffusely reflected sunlight or 'albedo' of different substances in nature.

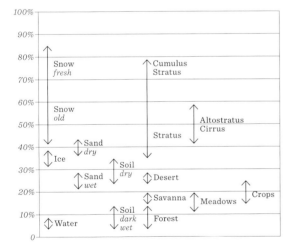

Opposite: A dropstone of oolitic limestone likely settled into this ancient seabed in Namibia after falling with some impact, perhaps from a volcanic event. The impact visibly deformed underlying beds and ejected a doubly-folded flap of the contemporary surface bed.

development of ever more detailed climate models generated with increasing computational power. The theory is now backed up by a raft of other studies that have made this a widely accepted event in Earth's long history. Seven hundred and fifteen million years ago the Earth, it seems, really was becoming a snowball, a whitening sphere hanging in space, a vision both beautiful and terrifying in equal measure. But as the planet continued to cool and the ice crept ever further down from the poles, it may have appeared that all was not lost. A band of warmth held out around the equator – a last refuge for life on the rapidly cooling globe – and the diminishing levels of carbon dioxide do not on their own explain what finally tipped the Earth over the edge. For the planet to turn into a snowball, our climate modelling strongly suggests that something else had needed to come into play, and we think that tipping point came about because of a very particular characteristic of ice.

Every surface on Earth has a particular relationship with sunlight – we call it the albedo of a surface, a quantitative description of the amount of the Sun's energy, the solar radiation, that is reflected or absorbed by any particular surface. Ice, unsurprisingly, has a high albedo, because it is very reflective of sunlight and therefore sends back out into space far more of the solar energy that strikes it compared to other surfaces, such as oceans or bare land. For this reason, the amount of ice on the surface of the planet has a direct connection with the climate, which means that any change to the Earth's ice coverage can have a profound effect.

Triggered by the plummeting carbon dioxide levels, the Earth's warming blanket thinned and brought on the advance of the ice, but as the white expanse spread, the planet became more reflective (higher in albedo), sending ever more of the Sun's heat back into space. This in turn further cooled the planet, allowing more ice to stretch across the surface until it eventually reached a tipping point whereby it became a runaway process: more ice, more reflection, more cooling, more ice. It was an unstoppable gallop that resulted in global glaciation on an unprecedented scale. Sea ice surged over almost all the remaining oceans, reaching a crushing kilometre thick in places, even at the equator. Then, finally, the last stretch of ocean succumbed to the ice, leaving the planet completely frozen – a white marble floating in the darkness of space.

It's almost impossible to imagine, but we think that for tens of millions of years pretty much our entire planet was just like this. The whole world from pole to pole was plunged into a deep ice age, wrapped in ice kilometres thick in places, choking the land and oceans. A global winter with no possible end in sight. Modelling suggests that temperatures reached -50 degrees Celsius at the poles, and it never got above freezing anywhere on the planet.

Unlike our warm world, this frozen planet would have been pretty sterile; it would have stressed or stalled the water cycle and there would have been little or no evaporation. Hence no rain, no snow and very few clouds – and if there were any they would have been clouds of carbon dioxide.

As the temperatures dropped and the ice enveloped the whole planet, it was a disaster for life. This was an incredibly hostile place. The ocean was cut off from the atmosphere, light levels under the ice plummeted and the flow of nutrients coming from the land slowed to a trickle. With this level of climatic trauma inflicted upon the Earth's ecosystems, those early food webs began to fall apart. Imprisoned by the ice, the predecessors of all animal life were left stranded, with a mass extinction being the only possible outcome from such hopeless circumstance.

This was no short, sharp, cold snap that life could easily ride out before the warm weather returned. The ice persisted, unending and unchanging, for year after year, millennium after millennium – a winter with no hope of spring.

OASIS IN THE ICE

'There is one habitat Snowball Earth scientists are particularly interested in – cryoconite holes. Sediment lands on the glacier surface, swept in by water and by wind, then radiation from the Sun warms it and melts the underlying ice almost vertically to make a pocket, open at the top and with meltwater at the bottom. The cryoconite holes provide the perfect living conditions for microorganisms because they have access to both the Sun and meltwater.'
Dr Jaz Millar, Bristol University

In the sterile conditions of the Snowball Earth, very little seemed to change on the icy surface of the planet for thousands of years. But looks can be deceiving, because even in this frozen stasis there *was* more dynamism to this frigid landscape than you might expect. Hidden amongst the endless sheets of ice there were still some isolated rock faces that remained ice-free, and it was here that the beginnings of a new home for life could begin to take shape. Not directly on the exposed rock itself – that environment was still too harsh to nurture life – but as the vicious arctic winds whipped around the bare rock they slowly stripped precious small amounts of rock dust away from the exposed surfaces. Carried on the wind, wherever that dark dust settled on the icy landscape it absorbed warmth from the Sun, creating something incredibly rare and precious on this frozen Earth – meltwater contained in structures known as cryoconites. And it was here in the most uninspiring of dusty puddles that the fate of our living Earth rested, because when it comes to life, where there is liquid water there is always hope.

Cryoconites are a miraculous oasis in a frozen wasteland, a feature that is found pockmarking ice sheets on Earth today just as it did millions of years ago. These small punctures in the surface of a glacier or ice sheet are created by the thermodynamic disruption that dark dust and ice bring to the ice when it settles in any significant amount. Due to its lower albedo compared to the surrounding ice, it absorbs a greater level of solar radiation, which leads to the ice melting beneath it, forming these strange, almost alien-looking cylindrical holes. The formation of these holes then serves as a trap for more material to accumulate inside – be it more of the small rock particles, soot or even bacteria. In this way cryoconites create a positive feedback mechanism and are a significant contributor to glacial surface melt.

Through this mechanism the holes can grow not just in size but in vitality. Extensive study of contemporary cryoconite holes has revealed detail of the extraordinary micro-ecosystems that can be formed in these dusty puddles. Mixed in with the mineral particles such as quartz, feldspar, mica and calcite, scientists have discovered algae, cyanobacteria, fungi and viruses all thriving. The holes can also harbour dormant seeds that have flown in from far away and even invertebrate life forms like rotifers and tardigrades. These sanctuaries for life are ultimately only transitory, and with time the layer of rock sediments becomes thicker, it isolates the underlying ice from heat and a reversed situation occurs – the ice around cryoconite melts faster, and a cone of ice with a cap of sediments is formed.

Despite their transitory nature we think these cryoconite holes played an extremely important role in the persistence of life through the Snowball Event that occurred 700 million years ago. And in part we know this because scientists have conducted experiments in the laboratory where they have exposed tiny puddles like this to the conditions found on Snowball Earth and the life inside has survived. It's a truly extraordinary thought to imagine our distant ancestors huddled away in these and other crooks and crevices across the icy globe. Toughing it out through the longest of long winters, life would have to endure a winter that stretched over 50 million years, without any sense of a nearing spring.

Below: Looking into a
cryoconite hole in a glacier
in Antarctica.

Right: A huge cryoconite
hole in Antarctica offers
tremendous habitat
potential for microbial life.

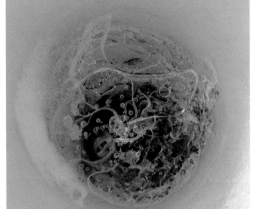

Right: A meltwater
channel with cryoconite
on the ground, filled on the
surface of the Gorner glacier
in Switzerland.

Cryoconites are a miraculous oasis in a frozen wasteland, a feature that is found pockmarking ice sheets on Earth today just as it did millions of years ago. Extensive study of contemporary cryoconite holes has revealed detail of the extraordinary micro-ecosystems that can be formed in these dusty puddles. One of the most well-known inhabitants of these oases of life is the tardigrade, also known as the water bear or the moss piglet. These microscopic animals are some of the hardiest on Earth, and have been found thriving at the tops of mountains, the bottom of the sea, tropical rainforests and the Antarctic. They have even survived exposure to outer space, in experiments conducted in 2007. Tardigrades are thought to be able to survive even complete global mass extinction events, gamma-ray bursts or large meteorite impacts. Some of them can withstand extremely cold temperatures down to 0.01 Kelvin (−273 degrees Celsius), close to absolute zero, so the species may have waited out Earth's Snowball Event in cryoconites.

RETURN OF THE FIRE

Locked in ice, the Snowball Earth seemed destined to stay in a frozen hell for eternity. With vast amounts of the Sun's heat being reflected back out into space and not a cloud in the sky to hold in even the merest scrap of warmth, temperatures at the equator reached as low as those we find in modern-day Antarctica, and as the world froze over it seemed as if there was little hope for the future of life on Earth.

But 650 million years ago there was still one chance of salvation, a route to recovery that had remained hidden deep beneath the ice throughout the frozen millennia. As the outer shell of the Earth shivered, the fire within the interior of our planet burned strong, and even though the ice had reached thicknesses of 1 kilometre in many places, not even that barricade could prevent the power of the Earth's volcanoes punching through the planet's frozen shell.

Now you might (understandably) think that at this point in our story the arrival of red-hot lava bursting through the ice at temperatures well over 1,000 degrees Celsius might have been the reason for the great thaw the planet was about to encounter. But despite the obvious and direct effect of a massive increase in volcanic activity at this time, we are pretty certain this was the trigger but not the actual smoking gun that led to life on Earth finally being set free from its icy grip.

Our leading theory to explain the big melt is that as each eruption tore through the ice-bound plains it released not just lava but a vast, invisible slug of volcanic gases that, crucially, included, among many other components, carbon dioxide. As we know, carbon dioxide is a potent greenhouse gas, and it was the stripping of this molecule from the atmosphere that initiated the global freeze in the first place. Now released back into this frozen world, its impact would be equally as transformative – but in the opposite direction. As concentrations of carbon dioxide increased, it would be the changing atmospheric composition that would trap ever more heat around the entire planet, warming the Earth and finally challenging the tyranny of the ice.

Below: An artist's impression of Snowball Earth, with no liquid water left on its surface.

Right: The Sangay volcano erupting in Morona Santiago, Ecuador, in 2020.

Far right: Cap carbonates are layers of distinctively textured carbonate rocks (either limestone or dolomite) that are created by ice ages and the subsequent super-greenhouse climate that follows them. These were formed in the Precambrian in South Australia.

'As each eruption tore through the ice-bound plains it released a mix of volcanic gases, including carbon dioxide. Each time it added to the atmosphere, leading to tiny increases in global temperature. As concentrations rose and built over millions of years the gas trapped more heat and the frozen planet began to thaw. Almost everybody agrees that, in geological terms, the melting, the deglaciation of Snowball Earth, was really fast. The dispute comes over whether fast means a matter of hundreds of years or maybe up to a million years.'
Professor Raymond Pierrehumbert, Oxford University

But this was no drip-by-drip melt. The latest evidence from the geological record suggests that the end of Snowball Earth was not a slow one. When it finally arrived, 635 million years ago, the thaw, it seems, came quickly, playing out over perhaps no more than a million years. Now that may sound like a long time, but as we have seen throughout our story, a million years is a blink of an eye in the Earth's 4.5-billion-year history. Trying to peer this far back into the past to create an accurate timeline of events is far from easy, but a group of scientists working in southern China's Yunnan province have uncovered evidence in rock samples there that seem to indicate that the Snowball Earth really did end in a flash.

To understand the evidence that this team, led by Shuhai Xiao from Virginia Tech, in Blacksburg, USA, have been unearthing, we have to delve a little deeper into the detail of the Snowball Earth event. From this distance of time it would appear that the almost complete glaciation event that we are calling Snowball Earth that started around 720 million years ago was a singular event ending with the big thaw around 635 million years ago, but scientists like Xiao and his team have been able to delve much further into that time and see that the Earth underwent a number of glaciations during the Neoproterozoic Era. We think there were at least three, or even possibly four, significant glaciations during the period known as the Cryogenian. Of these events the Sturtian was perhaps the most extreme, lasting from 717 to 650 million years ago, followed by the Marinoan, a shorter, sharper but no less severe part-Snowball Event that took place 650 to 635 million years ago.

Between these events the Earth experienced some sort of retreat in the ice, with glaciers extending and contracting in a series of seemingly rhythmic pulses, which at times allowed warmer waters to return to the Earth. For scientists like Xiao, the evidence of these warm seas can be seen in the deposition of a particular kind of rock formation called cap carbonates – unique deposits of limestone that were

'It is incredible to imagine that a global glaciation could have been undone in just a couple of thousand years. This is a geologically instantaneous amount of time.'
Dr Judy Pu, University of California, Santa Barbara

Above: Extreme monsoon flooding has submerged thousands of villages and huge swathes of farmland in northern India, Bangladesh and Nepal. In Bihar alone more than 515 people have perished and more than 16 million people in 7,972 villages have been affected. These floods – the worst in living memory – were caused by the sort of extreme weather that scientists have predicted will happen as a result of global warming.

Opposite: As global temperatures increase, Earth's weather is becoming more volatile. In Suesca, Colombia, a swimming lagoon has been abandoned after water levels reduced by 90% in the area in 2021 (top). In California, USA, in 2023, a sinkhole formed due to the powerful storms that have pounded much of the state's coast since the beginning of the year (bottom).

laid down as the conditions between ice ages that caused the oceans to act as a carbon sink, drawing in carbon dioxide and laying it down in stone. These limestone deposits are often found on top of volcanic layers of rock that hold direct evidence of the massive glaciers that sat there before the warm seas returned, and so it is at these boundaries that researchers are able to unpick and reveal the series of events and timings that pushed the Earth in and out of the deep freeze.

In some of the most accurate analysis of the transitions from glacial rock to cap carbonate, Shuhai Xiao and his team working in southern China were able to date the transitions using zircon crystal analysis. As one geologist commented, samples like this that are accurately datable are about 'as common as unicorns', and so the results that Xiao's team came up with provide a unique insight into this time period. So far they have revealed not only one of the most precise measurements for the end of the Sturtian ice age 658.8 million years ago (+-.5 million years) but the work has also enabled us to pinpoint the end of the Marinoan with astonishing accuracy. By comparing the age of volcanic rocks either side of the final cap carbonate layer of that epoch, this work has allowed us to peer into the past and see that the end of the Marinoan glaciation, the final part of the Snowball Event, ended rapidly and almost certainly globally. In geological terms, these two samples suggest that the melting event was an incredibly quick thaw playing out over no more than a million years.

This and the work of many other scientists investigating the Snowball Earth has provided a window into one of the most intense transitions in Earth's history, revealing how the Cryogenian Period finally came to an end. But this is not just a history lesson; it also reveals something profound about our planet today, because if the Earth can change that quickly, it inevitably challenges our complacency around the stability of our modern-day climate.

For most of human civilisation, our climate has been astonishingly stable and benign; it's only now that we are getting a taste of the stark reality that climate change can be rapid, brutal and runaway when certain tipping points are reached.

As we're learning to our cost, the very process of warming drives a cycle whereby more heat is captured by the planet. It's a dangerous loop, within which the temperature increase causes rising levels of evaporation to drive more water vapour into the atmosphere, and as this water vapour is itself a potent greenhouse gas, thus it increases the speed of warming. And that's just the beginning, because as more and more ice melts, the opposite of the albedo effect takes place, with the dark surfaces that are exposed as the ice retreats absorbing rather than reflecting warmth from the Sun and so adding more heat into the system.

These are just two of a host of consequences that can trigger warming; all of these changes come together to produce something that we call a positive feedback effect, whereby one tiny change can be magnified a hundredfold, creating a tipping point at which the changing climate is rapidly driven in a singular and extreme direction.

And 635 million years ago, the Earth was about to get a taste of just how powerful, violent and transformative that positive feedback process could be. As the grip of Snowball Earth finally weakened, what started as a trickle of melting ice became a torrent. Just as we are seeing today, as more of the oceans and land were set free, their dark surfaces sucked in warmth from the Sun, feeding more heat into the system and so making the melt unstoppable.

At the peak of the melt, we think that sea levels were rising two metres every decade – an unfathomably quick change that would have swallowed great chunks of coastline, endlessly re-shaping the newly exposed land. After a gruelling 50 million years in the deep freeze, the world was finally reborn, but this was no stable Eden, this was in climate terms a chaotic world, where the speed of change created a whole new set of opportunities and threats for the future of life.

The golden plover is a special bird. In Iceland, it is the equivalent of the cuckoo in more temperate climates, the harbinger of spring. People get so excited seeing the bird, they report it in the local press. It signifies an end of winter, the coming of warmth. And as for the plover itself, it has travelled a long way, perhaps all the way from Portugal, for this vast feast of fantastic food that's happening all around. One of the most wonderful things about our planet is that you can't consider pockets of life in isolation. They are all part of a global ecosystem full of interconnected, interdependent relationships, of dazzling scale and complexity.

A FRAGILE EDEN

Below: Cyanobacterial bloom in the Baltic Sea.

Below middle: A snapping turtle (*Chelydra serpentina*) swimming in a blue-green algal bloom. Microcystin produced by the algae can be toxic to wildlife and humans.

Below right: Dead fish suffocated by algal bloom from too many nutrients or not enough oxygen in water.

Snowball Earth was an unprecedented assault on the planet, the reverberations of which we are still living through today. We, and every living thing on the planet, are the survivors of that event – our ancestors were the ones who made it through the deep freeze and out the other side. But as those survivors emerged into a new era 630 million years ago, that ancient life experienced a very different spring from the one that greets us each time winter comes to an end. Both the simplest of bacteria and the most complex life forms that made it through Snowball Earth emerged into a world that was still intensely hostile. Eukaryotic life, the complex cells that would go on to develop into every plant, fungus and animal on Earth, had just made it through, the ultimate destination of complex ecosystems filled with complex multicellular organisms, an unimaginable destination from this barren starting point.

But in the shattered remains of the great freeze, it was this very disputation that would set the stage for a revolution, the chance for complex life to not just survive but finally break free and redefine its relationship with the planet.

We often assume that life progresses in a straight line, steadily becoming more complex, more advanced, but in reality it's rarely like that. Great leaps in evolution often seem to rely upon great shocks to the status quo. It is, more often than not, the deepest of disruption that flips life onto another path, after which that life and the world in which it lives are never the same again. Snowball Earth was precisely one of these definitive moments.

If we could survey this scene of destruction 630 million years ago, we would see a planet reeling from the after-effects of the great thaw. The ice of Snowball Earth had shredded the surface of the planet, ripping up millions upon millions of tonnes of rock and dumping it all into the newly freed oceans. And as we've seen before in this chapter with the breakup of the supercontinent Rodinia, where minerals flow, life can flourish. So this sudden flood of resources into the oceans had a profound

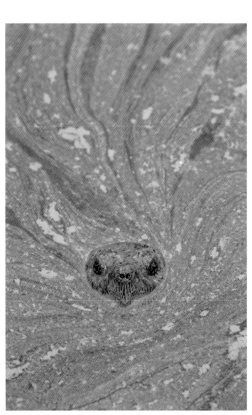

'Our Earth finally became the stage that ushered in a new age – one of complex life that would eventually reach the scale, beauty and complexity that we see today.'
Professor Peter Girguis,
Harvard University

effect on life, causing single-celled algae and bacteria to flourish throughout the waters of the world. These vivid algal blooms would have been so large they would have been visible from space, stretching across entire oceans and forming the basis for a new foundation upon which life could build. This was growth on such a staggering scale that it had the power to change the entire ecosystem of Earth. From the surface to the depths of the ocean, the rain of dying algae and bacteria created a vast graveyard of dead, organic matter, a flood of food that along with the mineral resources pouring off the land supercharged the ocean habitat. In this world of plenty, bacteria could gorge themselves on the dead matter, reprocessing it and releasing a steady stream of organic nutrients back into the water column, and for the first time in millions of years our ancient multicellular ancestors had a reliable, consistent flow of resources, not just from the land but from life itself.

At this moment, for the first time ever, life and the planet became complementary cogs, spinning in a great nutrient recycling machine, which provided life with exactly what it needs in quantities that it had never had before. And as a consequence life exploded; after around half a billion years of struggle, Earth's fledgling spring was set to become the most glorious of summers.

Below: A modern-day example of a glorious bloom of life as juvenile crabs return from the sea on Christmas Island, Australia.

A GOOD LUNCH

To uncover just how dramatic the explosion of life was after the Snowball Earth would take the work (and good fortune) of a young Australian geologist called Reg Sprigg back in the 1940s. Sprigg was something of a star student, becoming the youngest Fellow of the Royal Society of South Australia at just 17 years old. In the early part of his career, he worked for the South Australian Geological Survey mapping uranium mines in the Australian outback as part of the allies' attempt to secure a reliable source of the uranium for the top secret Manhattan Project during the Second World War.

After the war, Sprigg continued his work for the South Australian government but this time on much less secret and much more mundane missions. His task was to explore whether old abandoned mines 650 kilometres north of Adelaide in part of the Flinders Range called the Ediacara Hills could be transformed with the new mining technology of the time into a profitable venture. The story goes that Sprigg was having his lunch after a long morning's work, when he realised the very flat, weathered sandstone rocks he was sitting on looked remarkably like the remains of an ancient seafloor. Sprigg had grown up near the beach in Adelaide and so had spent much of his childhood hunting for fossils that he often shared with academics at the city university. With his eye firmly caught by these ancient-looking rocks, he began inspecting them, and even though palaeontologists back home had told him again and again that flipping sandstone rocks is unlikely to reveal any fossils, his curiosity led him to turn one of the slabs over and in doing so he would ultimately change our understanding of life on Earth.

Opposite top: A flatworm (*Dickinsonia costata*) fossil from the Ediacaran biota, Precambrian Period.

Opposite below left: *Spriggina* is a genus of Ediacaran bilaterian animals named for Reg Sprigg.

Opposite below right: A fossilised jellyfish (*Mawsonites spriggi*), a specimen of the Ediacaran fauna from the Precambrian of Australia. It is thought to have lived around 700 million years ago.

Right: Wilpena Pound, a natural amphitheatre in the northern central part of South Australia's largest mountain range, the Flinders Ranges.

LIFE IN THE CAMBRIAN
The evolution of different animal phyla over the course of the Cambrian Period based on fossils discovered.

On the underside of one of these rocks Sprigg found a distinctive circular fossil that looked remarkably like some kind of jellyfish. Other fossils would quickly follow, and dating suggested this complex life was over 600 million years old, so like any good geologist armed with what seemed like a major discovery he attempted to publish his findings in *Nature*. The paper, however, was rejected, as was the significance of the Ediacara Hills as a transformative site in our understanding of life on Earth. This was because at that time it was unthinkable that such complex life could be discovered in rocks that old, and the overwhelming consensus was that complex life had first emerged on Earth around 540 million years ago in an event known as the Cambrian Explosion, and that life before this was nothing more than bacteria and other microorganisms. And so the fossils Sprigg had found in those rocks that dated back over 600 million years were dismissed as inorganic remnants rather than the remains of living creatures. It would take more than a decade before a series of other discoveries (including another moment of palaeontological serendipity, this time in Charnwood Forest in Leicestershire, in England, when three schoolboys found an ancient fossil while walking in the woods) would allow palaeontologist Martin Glaessner to link Sprigg's jellyfish from the Ediacara Hills with the plant-like fossil from Charnwood Forest and a series of other frond-like fossils discovered in Namibia earlier in the century. Glaessner suggested that these findings were clear evidence of complex life existing long before the Cambrian Explosion. Science had opened a window into a whole new age of life on Earth, an age that was formally named the Ediacaran Period after the location that provided that first, fundamental, breakthrough.

Above: A fossilised sea pen (*Charnia masoni*), another Ediacaran animal, found in Charnwood Forest, England.

Right: *Rangea schneiderhoehni*, a frond-like Ediacaran fossil discovered in Aus, Namibia. *Rangea* was the first complex Precambrian macrofossil named and described anywhere in the world.

CAMBRIAN EXPLOSION

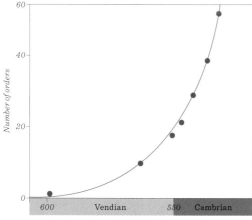

Geologic time (millions of years ago)

Above: A bloom of
Ediacaran jellyfish preserved
in the fossil record of the
Ediacara Hills.

GARDEN OF EDIACARA

So this was the world that the Snowball Earth gave birth to 600 million years ago. In shallow coastal waters across the planet, life was blossoming into myriad new and strange forms. For the first time in Earth's 4-billion-year history, large multicellular life was widespread. The direct descendants of those first tiny eukaryotic pioneers that had hunkered down and survived, generation after generation, through millions of years of deep freeze, had now begun to evolve and flourish. Breaking free of the microscopic world, these were creatures of a size and complexity that would be recognisable to us, all living within and supported by a web of life just as intricate and remarkable as anything on modern-day Earth. In the last fifty years, fossil evidence of this 'Garden of Ediacara' has been discovered all over the world – from the west coast of Mexico to the white coast of Russia. In locations like this, evidence of once-thriving ecosystems has been uncovered, revealing a vast array of strange, morphological characteristics, varying in size from millimetres to metres, with body plans that range from the complexity of a blob to intricate symmetrical structures, with differentiated cell systems and complex behaviour.

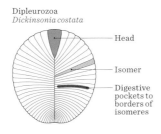

Vendiamorpha
Vendia sokolovi

Isomer

Digestive
system

Dipleurozoa
Dickinsonia costata

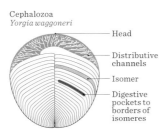

Head

Isomer

Digestive
pockets to
borders of
isomeres

**COMPARING THE
PROARTICULATA**
Proarticulata is a proposed
phylum of extinct, bilaterally
symmetrical animals known
from fossils found in the
Ediacaran marine deposits
dating from approximately 567
to 550 million years ago. Three
examples of this phylum are
Vendia sokolovi, *Dickinsonia
costata* and *Yorgia waggoneri*.

Cephalozoa
Yorgia waggoneri

Head

Distributive
channels

Isomer

Digestive
pockets to
borders of
isomeres

This was a world filled with alien, unrecognisable life forms that we are still trying to characterise and comprehend. Life forms like *Dickinsonia*, an organism that has been described as a 'segmented bath mat', were no more than just a few millimetres thick but could grow to nearly 1.5 metres in length, living out their life on the Ediacaran seafloor. This organism is so alien to us that there has been much debate over what category of life form it actually was – protozoa, plant or animal? Only recently has biochemical analysis of fat molecules discovered remaining on the surface of the fossil allowed us to decide with some certainty that the *Dickinsonia* is almost definitely part of the animal (Metazoan) kingdom.

Equally strange and no less mysterious is *Ikaria wariootia*, the oldest-known animal to have a mouth and gut (therefore making this the oldest arsehole ever discovered). These tiny creatures, just a millimetre or so wide, burrowed their way through the seafloor leaving characteristic tiny, twisting tunnels that are one of the most commonly found characteristics in Ediacaran sediments. But it was only when scientists from the University of California, Riverside, used a 3D laser scanner to visualise the burrows that they found hundreds of tiny blobs in the

Left: An illustration of
the marine life found in
seas between 580 million
and 560 million years ago,
in the Ediacaran Period.
Fossils of all the creatures
shown here have been
found in Precambrian
rocks worldwide: jellyfish;
Mawsonites spriggi (blue);
Kimberella quadrata (bullet-
shaped); sea pens; flatworms
(on seabed); and algae (dark
green tufts).

Above: An illustration of
deep-water Ediacaran
fauna that lived 635 to
539 million years ago. This
period was characterised by
the emergence of the first
multicellular animals such as
frond-shaped rangeomorphs
(centre) and bilaterian proto-
molluscs (left).

Left: A palaeontologist works with *Dickinsonia* fossils in the Ediacara Hills, South Australia.

Below: Hollows left by *Dickinsonia* specimens in seafloor mats. *Dickinsonia* were flatworms that grew by adding body segments up to 80.

Above: Over 550 million years ago in South Australia, *Ikaria wariootia* was a worm-like bilaterian animal which measured between 2 and 7mm long and burrowed in sand on the ocean floor.

tunnels. Close inspection revealed these to be organisms no bigger than a grain of rice, capable of not just exploring their landscape but also scavenging for food within it.

Everything, it seems, in the Garden of Ediacara could turn our understanding of life on its head. In Newfoundland, Canada, ancient rocks have revealed some of the oldest Ediacaran species we have ever discovered, which at first glance look like an ancient plant with a structure that resembles the fronds of a fern. These ancient organisms are in fact not a plant but an animal, part of a collection of Ediacaran life known as the *frondose*. Many of these creatures built from miniature branches have no reference point on the modern Earth. Organisms like *Charnia masoni*, a frondose animal that resembled a palm branch 'stuck to the seafloor' and that was made up of a fractal structure of branches and ever-smaller sub branches, is unlike any animal living today and has a unique structure not seen in any plant or algae either. A cousin of *Charnia* also discovered in Newfoundland is an animal belonging to the genus *Fractofusus* that looks like a mound, with fronds shaped in a disc on the seafloor. *Fractofusi* have recently been discovered to live in 'family' groups, with larger ones surrounded by clusters of smaller discs. These *fractosfusus* 'children' are thought to be a product of a reproductive mechanism that sounds as weird as the animal they came from, produced by the parent organism sending out long threads up to 4 metres long, upon which the offspring developed.

Exactly when the first truly complex living creatures of the Garden of Ediacara died out is still unknown, as is whether they lingered on to help drive the next extraordinary evolutionary explosion of life that defined the Cambrian Period. But what is certain is that all the strange life forms that made up this fauna are now long gone; lost, as so many species have been, to the depths of time. Their soft bodies only rarely preserved in the most precious of fossil beds, they have given us a glimpse of the moment in the Earth's history when complex life took its first great leap – not just in creating living entities, but also in creating an array of life forms that could form biologically rich ecosystems that became self-sustaining and supportive of food chains that drove a new wealth of behaviours. From this point in time there was no turning back, thanks to Snowball Earth lighting the fuse – this was the foundation upon which everything we recognise as a living Earth would be built.

Snowball Earth was an astonishing period during which life, the land and the climate were in conflict like never before. After the Earth's great freeze, the stage was set for evolution to run riot and give us the fabulous scale, the astonishing diversity and the remarkable beauty that we see among life on Earth today. But before that could happen another great hurdle would have to be overcome. Life would have to conquer not just the oceans but the land too, and to do that would require the rise of perhaps the most successful life form on Earth. To understand how plant life, against all odds, eventually made its way onto dry land we have to leave the Garden of Ediacara behind and once again return to the early Earth, where the story of our green planet begins.

Left: A seascape in the shallow water of a coral reef in the Caribbean Sea – perhaps a modern version of the Ediacaran garden ecosystem.

Above left: Ancient animals such as sea fans, sponges and soft corals thrive in shallow seas off the coast of Belize today.

GREEN

'No matter how hard we try, we can't
escape the fact that this is a plant planet.'
*Dr Kirk Johnson, Smithsonian National
Museum of Natural History*

GREEN
EARTH

When viewed from space, the character of our planet is defined by the complexity of its colours – the deep blue of the oceans shimmering across vast swathes of its surface, yellow seas of sand that shift and flow across the desert lands, and white clouds that trace out the motion of our atmosphere as it dances around the planet. But perhaps more than any other there is one colour that reveals the depth of our planet's preciousness, a colour that clearly announces that this is a living, breathing planet.

Ours is a predominantly green planet – a world where life shouts out its existence to the Universe in a single powerful hue. From space we can see this in the great green brushstrokes of the forests of the Amazon, the 700,000 square miles of lush foliage in the Congo Basin, or closer to home in the rolling verdant fields of the United Kingdom. But it's only when you come down to Earth that the true dominance of this colour begins to make itself clear. So much of our planet is covered in a rich green carpet of vegetation, so much of our everyday view of the world is peppered with green, that at times we miss just how successful, how dominant and how pervasive plant life is on the surface of our planet. In every corner, crook and crevice, plants make a home. Just take a look around you now, wherever you are, on whatever continent and in whatever season, and the chances

Below: The International Space Station affords sensational views of our planet and the tapestry of land and oceans.

are you are just a few metres away from something that is green and living. Whether it's a grand forest, a square of parkland, a scrap of grass or just a few weeds growing in the cracks of a pavement, it is astonishing when you take a moment to reflect on how well plant life is able to adapt and thrive in almost every environment on Earth.

It is the ultimate success story of life on our planet. For millions upon millions of years the Earth has been green, the only world in our solar system – and perhaps even the only world in our galaxy – that has found a way to tap into the light of a star and live off the endless stream of energy that comes with that capability. And with it has come the foundation of the complex web of life that we experience; the success of plants has shaped the land, the air and ultimately the endless diversity we see here. But it didn't have to be this way. The wealth of plant life that we see around us today is far from an inevitable consequence of life existing on our planet. In fact, when life began around 4 billion years ago, this green conquest of the planet looked anything but possible.

Above and left: From the giant water lily pads (*Victoria amazonica*) in the Amazon to lush woodlands (middle) and rainforests (bottom), the wealth of plant life all over Earth is evident in our green planet.

Packed with an astonishing abundance of biodiversity, the lush green carpet that covers the Nā Pali coast on Kaua'i, Hawaii, stretches all the way down to the sea. And it smells so good – that rich organic mix of the soil that gives our planet its name. In this area alone there are probably 100 different species of plants, each with its own unique story.

Take a look at this little beauty; this is *Lysimachia glutinosa*. It may not be the most glamorous plant in the world, but its claim to fame is that it grows here and only here, only on this one side of this one small island. This makes it sound very fragile, very vulnerable, but that couldn't be further from the truth. Because in the story of this plant – and, indeed, all plants – lies the story of our Earth.

THE HADEAN EON

H alf a billion years after Earth was born, the planet was in a period of momentous transition, as the first of its four great eons, the Hadean Eon, was coming to a close. In 500 million years the Earth developed from the newly formed, volatile planet of its birth, with its hot molten surface and a toxic atmosphere rich in methane, ammonia and the noble gas neon, and lacking free oxygen. It survived being hit by an object about the size of Mars, named Theia, and that near-cataclysmic impact saw the Moon form in its night skies from the resulting debris. Despite the bombardment of an endless battery of meteorites, the Earth not only survived, but its surface was transformed from a lava world to a water world . Covered by an endless ocean, the only landmasses that broke through the surface of these waters were a few remote volcanic islands, transitory formations that soon crumbled and disappeared again beneath the waves. Four billion years ago, any dry land on Earth would have been no place to call home; recycled repeatedly by explosive volcanic eruptions. It was an era when the heat from the Earth's formation and the abundance of rapidly decaying radioactive elements meant that volcanic activity was far more frequent and violent than it is today. This was a world with no place for plants; there was no stable land, no surface to terraform. Instead, the

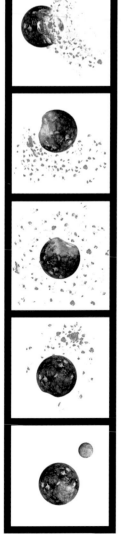

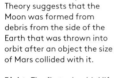

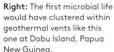

Left: The Giant Impact Theory suggests that the Moon was formed from debris from the side of the Earth that was thrown into orbit after an object the size of Mars collided with it.

Right: The first microbial life would have clustered within geothermal vents like this one at Dobu Island, Papua New Guinea.

OPPOSITE:
Top left: The gravitational pull of the Moon creates tides in the Earth's oceans, helping to power life in the water world.

Top right: Combined infrared and optical satellite image of volcanoes on the island of Java, Indonesia. Merapi (right) is the most active of these volcanoes, having erupted 68 times since 1548.

Bottom: Framed by components of the International Space Station, the Full Moon and the atmosphere are visible in this view above Earth's horizon.

'A watery environment is lovely if you're a plant. It's like being surrounded by a bath of your food. You've got nutrients available to you, there's no danger of drying out and you don't have to worry about gravity or soil.'
Professor Katie Field,
Sheffield University

ancestors of all life, including the descendants of every plant that we see all around us on the surface of our planet today could be found growing in the depths of the oceans. Sheltered inside the walls of geothermal vents, the first microbial life clustered there and, dependent on the energy of the Earth for its survival, this life was stuck in the depths. Our planet may have had its moment of genesis, but any green Garden of Eden was still a long way off, with the conquest of the land being essentially impossible at that point.

So the question is – and it's a big question – how did life forge a permanent base on the land? Because if the Earth's only trick when it came to land building had been volcanism, it's very likely that life would never have made it out of the ocean.

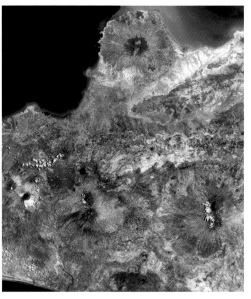

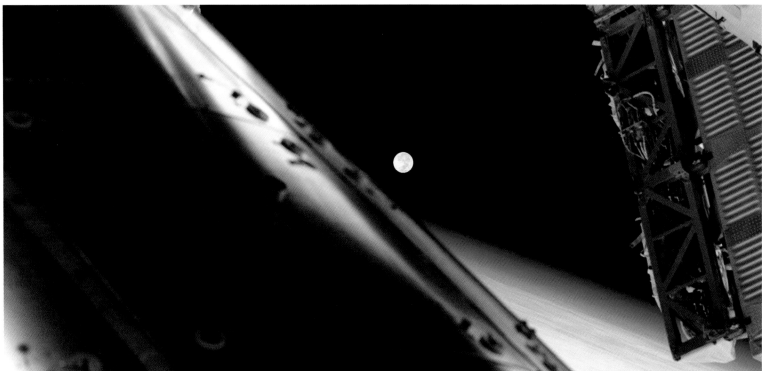

Above: The layers of the Earth's core, from the mantle through to its burning centre.

To understand what changed, we first need to comprehend how our planet functions today. Beneath us, the internal structure of the Earth can be divided into four distinct layers. The inner core at its centre is a solid ball of iron and nickel with a radius of around 1,220 kilometres, which is about three-quarters the size of the Moon. Surrounding this is a fluid layer of the same two elements, approximately 2,400 kilometres thick, which is known as the outer core. It's the flow of molten iron around this outer core that generates the Earth's magnetic field and protects us from the violence of the solar winds that travel at speeds of up to 1 million miles per hour. Above this is the planet's thickest layer, the mantle, which is almost 3,000 kilometres thick and composed of silicate rocks rich in iron and magnesium, and which becomes less and less malleable the closer it gets to the surface. Above this is the Earth's outer layer, the crust, which ranges from 5 to 70 kilometres deep across the surface of the planet – the thinnest parts are the oceanic crust composed of igneous rock like basalt and the thickest parts are the continental crust made of granite. Together, the Earth's crust and uppermost mantle function in a layer known as the lithosphere, a rigid shell that acts to remodel and remake the Earth's surface through a seemingly neverending process of remodelling that we call tectonics (from the Greek word *tektonikos*, meaning to build).

The reason the Earth's surface is so dynamic compared to the other rocky worlds in our solar system is because of the way the lithosphere is structured. Mars was once almost certainly a dynamic geological world, but today it has a 'stagnant lid', a lithosphere that is still and stationary. There are no earthquakes on Mars, no new

'Rocks breathed life into the land, but then life altered the rocks and changed them and diversified and enriched our planet in so many ways.'
Dr Robert Hazen,
Carnegie Institution for Science

mountains – it is as if the planet's surface is frozen in time. In contrast, on Earth we see dynamism on the surface of our planet everywhere we look, and the reason for this, the driving force behind it all, is that our lithosphere is cracked into seven large plates, along with a number of smaller ones, that all float on top of the mantle in a continuous continental dance. The movement of these plates may be incredibly slow, but the forces that the movement of these continental plates generates is enough to move mountains, to shake the Earth. Crucially in the context of the story of our planet, it was the creation of the continental plates that was the driving force behind the transformation of our planet from a water world into a terrestrial world, forming landmasses stable enough to allow plant life to begin its slow journey out of the water onto solid ground.

Above: Coral reefs exposed by an earthquake in the Pacific that lifted the island of Ranongga by 3 metres.

THE MESO-ARCHEAN ERA

ASTEROIDS AND COMETS

NASA's Wide-field Infrared Survey Explorer (WISE) surveys asteroids and comets in the Solar System for the project NEOWISE. This data was collected over eight months. The perspective is looking down from high above Earth's North Pole. The planets are shown as dark blue dots, and their orbits are the dashed lines. Earth is the third blue dot from the centre. The big mass of black dots consists of asteroids in the main belt, between Mars and Jupiter.

The green dots show the known near-Earth objects, both asteroids and comets, that NEOWISE has spotted so far. New-found near-Earth objects are shown as red dots. The turquoise squares are comets observed by the mission so far, and the yellow squares are newly discovered comets.

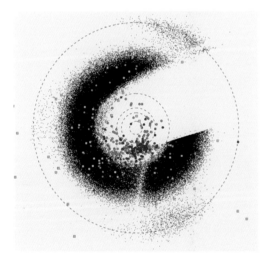

Opposite: The Geminid meteor shower over the Very Large Array Radio Telescope in New Mexico, USA.

It all began to shift almost three and a half billion years ago. On the surface, the Earth was still a water world, with only the most transient of landmasses punctuating the endless ocean. Back then, our planet's crust was one continuous solid layer, peppered with the occasional crack, but nothing like the complex interaction of continental plates we see today. But this was about to change.

Nobody knows for certain the series of events that led to the fracturing of the Earth's crust, a transformation that would change the course of the planet's history for ever, but we now think that one of the major turning points came in the form of a celestial intervention.

In its first billion years of existence the Earth had been continually bombarded with asteroids. This was a chaotic solar system, yet to settle down from its violent formation. The frequency of asteroid impacts was far higher than it is today; there were periods of calm, but compared to now these were few and far between. So as primordial life struggled to maintain its tenuous foothold in the oceans three and a half billion years ago, it also had to deal with the planet being continually bombarded by giant asteroids, bigger than at any other time in Earth's history. Some of these are thought to have been as large as 70 kilometres in diameter – more than eight times the size of the one believed to have collided with Earth, and which caused the extinction of the dinosaurs over 66 million years ago.

The consequence of this barrage is almost unimaginable; any early life forms in their direct path would have been instantly lost to oblivion. But for all their destructive force we now think these repeated impacts were instrumental in transforming this early water world into something very different.

This theory suggests that the force of a meteorite impact or impacts may have been instrumental in fracturing Earth's crust into those giant subterranean plates. Punching right through the crust and into the mantle, the vast amounts of energy from an impact of this scale would have melted the mantle locally, setting in motion a series of deep convection currents that would begin to drag the newly fractured plates along with them, thus starting the process of tectonic movement. Once these newly formed plates began to move, they would have impacted the whole planet, colliding and crashing against each other and in doing so spreading the process of plate tectonics.

Tantalising as this theory is, it is not the only hypothesis that has been put forward as the trigger that put tectonic plates into action. Another theory suggests that hotspots on the Earth's core could have triggered the change, forcing plumes of hot mantle to rise up from deep within the Earth, which would have caused the crust to weaken, crack and sink, creating a self-sustaining cycle that eventually turned into plate tectonics. With or without asteroid impacts or hotspots at the centre of the Earth, what we are certain about is that around two and a half billion years ago plate tectonics kicked into action, initiating the most powerful land-building force that Earth has ever known, and from that point the destiny of the planet changed forever.

These plate tectonics had such a dramatic effect on the nature of the Earth's surface because they initiated a vast recycling system that over geological time caused the crust to be endlessly renewed, which also created a new type of rock that would transform the destiny of life on Earth. This may all seem like a long way away from our story of plants and our green Earth, but the creation of the tectonic plates, be it by meteorite strike or some other process, was instrumental in the history of life on Earth.

SPINNING PLATES

Our journey to understanding that the Earth was quite literally moving beneath our feet was, as with many of the greatest scientific discoveries, both long and far from straightforward. For centuries many great minds speculated on the origin of the continents and the forces that lay behind their creation. In the late fifteenth and early sixteenth centuries, Leonardo da Vinci was one of the first to hypothesise that the grandest of mountain ranges had risen from the oceans, a theory that, however unfathomable, could explain why the remnants of ancient marine life had been found in the rock of alpine peaks.

But it wasn't until the beginning of the twentieth century that a mechanism to explain these earlier theories began to emerge, which had its origins in the work of a German geologist and climatologist, Alfred Wegener, the father of plate tectonics. Wegener lived and worked in the climatology field all his life, surviving three expeditions across the harsh terrain of Greenland until he eventually succumbed to the extremes of this unforgiving environment during his fourth expedition to the Arctic icecap in 1930. Wegener now lies buried deep beneath the snow and ice he spent decades exploring. But despite all of his work in this most extreme of environments, it wasn't here that Wegener first stumbled on his revolutionary idea of continental drift; the first inkling of this theory came to him from the simple observation of nothing more complex than an atlas of the world.

Top: Leonardo da Vinci's rendition of a storm over a mountain valley, circa 1500.

Above: Alfred Wegener, the German geophysicist and polar researcher at the base camp for Johan Koch's 1912–13 Greenland expedition (left), and on his own expedition in 1930 (right).

Sometimes the world can be blind to the blatantly obvious. Thousands upon thousands of people would have looked at the same map of the world as Wegener did towards the end of the first decade of the new century, and they too would not have seen what was staring them in the face. Wegener was the first to notice that all the continents of the Earth appear to slot together like a gigantic global jigsaw puzzle: South America connects with Africa, North America with Europe, Antarctica and Australia are a perfect fit, and India and Madagascar interlock precisely with the tip of South Africa. Wegener hypothesised that the reason for all of this

interconnectivity was because these landmasses had in fact been joined together in an ancient supercontinent that he called in German the Urkontinent (primal continent), and in Greek, Pangaea.

Analysis of the geology and fossil records from both sides of the Atlantic provided direct evidence that these distant lands were of the same origin, which resulted in his landmark publication *The Origin of Continents and Oceans* in 1915. But providing a mechanism to explain how this great fracturing and movement could have happened was not so easy, and without a mechanism to explain it the reaction to Wegener's work was at best sceptical and in the main dismissive. For years his theory languished under a constant bombardment of criticism, and with no ability to explain the forces that drove his theory of continental drift Wegener would not live long enough to see his notion vindicated.

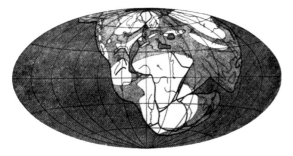

Top: The Tibetan Plateau was formed when the north-moving Indo-Australian Plate, moving at about 15cm per year, collided with the Eurasian Plate about 70 million years ago.

Above: Alfred Wegener's theory of the supercontinent Pangaea and how it split to form the present-day continents, from his *The Origin of Continents and Oceans*, 1915.

Right: The Himalayas, seen from the space shuttle Discovery during Mission STS-29 in 1989.

Right: The first scientific map of the Atlantic Ocean floor, created by Bruce Heezen and Marie Tharp in 1977. Tharp and Heezen's work revealed the presence of a continuous rift valley along the axis of the Mid-Atlantic Ridge, leading to acceptance of the theories of plate tectonics and continental drift.

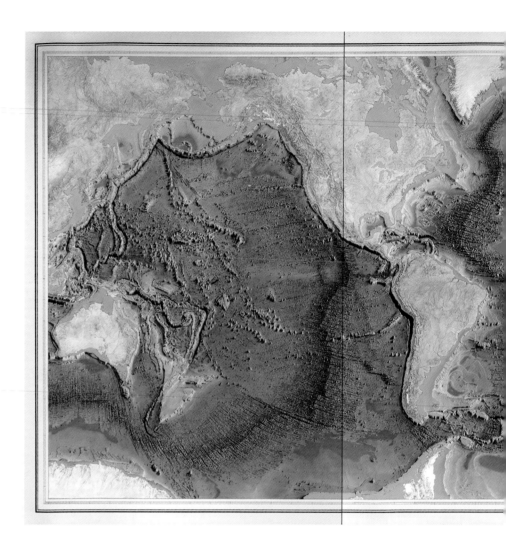

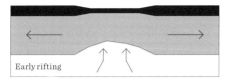

Early rifting

Continued thinning of crust and mantle lithosphere

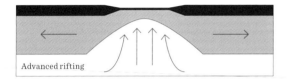

Advanced rifting

Passive margin   New oceanic crust created at mid-ocean ridge

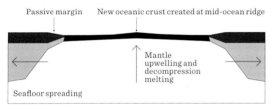

Mantle upwelling and decompression melting

Seafloor spreading

SEAFLOOR SPREADING

SUBDUCTION

Subduction occurs where two tectonic plates meet. The lithosphere of one plate is driven under the other, where it is melted and recycled in the mantle. The regions in which this occurs are called subduction zones, where volcanoes and earthquakes are common. This process creates mountains and trenches, and subduction in the past formed most of the Earth's continental crust.

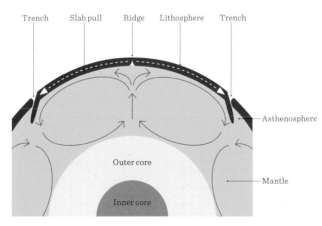

Trench   Slab pull   Ridge   Lithosphere   Trench

Asthenosphere

Outer core

Mantle

Inner core

It wasn't until the technological arms race driven by the Second World War and the invention of radar technology that our ability to map the planet, both above and below the oceans, began to reveal evidence that would ultimately vindicate Wegener's work and turn his fringe theory into the consensus scientific theory that plate tectonics is today.

In the years after the war, scientists were able to begin using the newly declassified radar technology to explore the alien landscapes beneath the ocean depths. A team led by the oceanographer Maurice Ewing and another team from the Woods Hole Oceanographic Institution, USA, used the research vessel *Atlantis* to explore the topography of the Atlantic Ocean. They were able to discover that the seafloor of the central Atlantic was not made of granite like the continental landmasses, but instead the crust in this mid-Atlantic ridge was comprised of basalt and was much thinner. The discovery of similar ocean ridges across the globe all pointed to one conclusion – that deep beneath the ocean a new floor was being created in a process known as 'seafloor spreading'.

This discovery was direct proof of the dynamism of the Earth's crust. Far from frozen in time, it was clear from the discovery of the mid-ocean ridges that the Earth itself was endlessly changing. But this discovery also created another mystery: if the seafloor was expanding then it was reasonable to suggest that the Earth must be expanding too, a theory that, as unlikely as it sounds, was actually

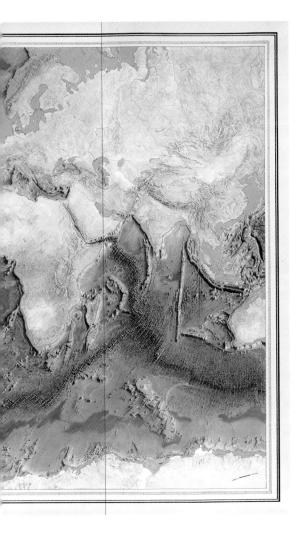

Above right: A new eruption
of Mount Etna in Catania,
Italy, in 2022.

a well-respected hypothesis right up until the 1960s. To dismantle the 'expanding
Earth' hypothesis would require an explanation of how the new crust that was
continuously adding along the oceanic ridges could be created without increasing
the size of the planet. The answer lay in another great geological feature that had
been recently discovered and detailed deep down on the seafloor. The new crust
that was forming was in fact causing the excess old crust to be driven through the
grandest of recycling processes, along trenches deep beneath the planet's oceans.
It was here in the oceanic trenches that a process known as subduction was
discovered. The oceanic crust driven in a great conveyor belt of new rock was, after
millions of years, consumed back into the Earth at the same rate as the new seafloor
was being created. And where were these trenches situated? The answer would
provide the evidence to prove Alfred Wegener's long-held theory: subduction occurs
at the boundaries of the great continental plates, where the pieces of the shattered
outer shell of the Earth meet each other and a heavier plate is thrust beneath the
second plate, where it sinks into the mantle. Take the quickest of glances at a map
of these subduction zones around the planet and it is clear to see that these are

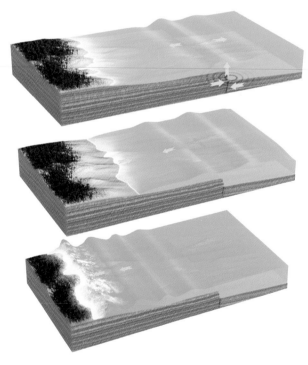

Above: A tsunami forms after an underwater earthquake (top). The seabed has moved upwards and the displacement of the water produces a series of powerful waves (concentric circles). These waves move rapidly (hundreds of kilometres an hour) in the open sea. As they approach a shoreline, they slow down and bunch up, sometimes towering tens of metres above the original sea level.

Top: Early-morning mist below Mount Fuji, reflected in Lake Kawaguchi, Japan.

also the locations of intense geological activity. Iconic volcanoes like Mount Etna, Mount Fuji and Mount St Helens all lie in subduction zones, as do areas prone to earthquakes and the creation of tsunamis. In the last 100 years, nine out of ten of the biggest earthquakes have been caused by the motions of tectonic plates in subduction zones, as well as catastrophic events like the Boxing Day tsunami of 2004 in northern Sumatra, Indonesia.

Although events like this are immensely destructive on a human scale, on a geological scale the process of subduction at the oceanic trenches is actually one of creation. As one plate gets pushed under another, vast quantities of basalt, seawater and sediment are flushed down into what is the equivalent of a highly charged planetary pressure cooker. And it's here where this particular combination of geological ingredients undergoes a profound transformation and, after a period of slow cooling closer to the surface, finally emerges as a new type of rock, an offspring of basalt called granite.

The creation of granite through the process of subduction might at first sound like just another of the Earth's multitude of geological processes, but its creation through the movement of tectonic plates has been fundamental in the shaping of our planet, both now and in the past. And that's because granite has a very particular characteristic compared to basalt that makes this hard, heavy, unforgiving rock a game changer in terms of the surface of the Earth. Counterintuitively, granite is actually 10 per cent less dense than basalt, so this means that when it's formed deep down inside the Earth, it naturally rises to the surface and, just like an iceberg on an ocean, it floats. This additional buoyancy has built the continental crust upon which you are now sitting, a crust that, because it is floating on top of the continental plates, is actually far older than the endlessly recycled oceanic crust. It's this process of tectonic collision forcing the basaltic crust down and pushing the newly formed granite up that creates the terrestrial world around us in all its wonder – from vast mountain ranges to meandering river valleys. Across the planet and over the course of geological time, more and more granite has accumulated on

'With the advent of plate tectonics comes an entirely new kind of rock – granite – which could allow life to get a foothold. Granite has a low density; we all understand density because we put ice cubes in drinks, and when we do about ten per cent of it sticks above the water, because ice is ten per cent less dense than water. And in the same way granite is ten per cent less dense than basalt, so ten per cent of the granite sticks up above basalt. Thus Earth's first landmasses are born.'
Dr Robert Hazen,
Carnegie Institution for Science

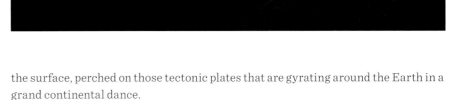

Above right: Astronauts train in a lava tube as part of a course simulating geological fieldwork on a planetary mission to Mars. Lava tubes on planets other than Earth may contain subsurface water, and are potential candidates in the search for extraterrestrial microbial life.

Below: Basalt on Earth (Svartifoss, Iceland, left), the Moon (middle) and Mars (right).

the surface, perched on those tectonic plates that are gyrating around the Earth in a grand continental dance.

We don't just think this theory, we've seen it in action. In the space of just a few years we went from plate tectonics being a contentious and unproven theory to being able to directly observe the impact of these plates on the positions of the continents in real time, thanks to the extraordinary work of one unusual satellite and the mission team behind it.

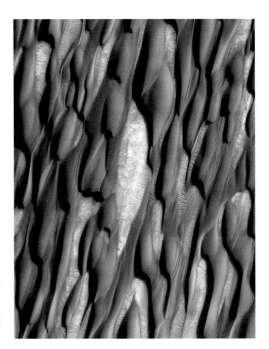

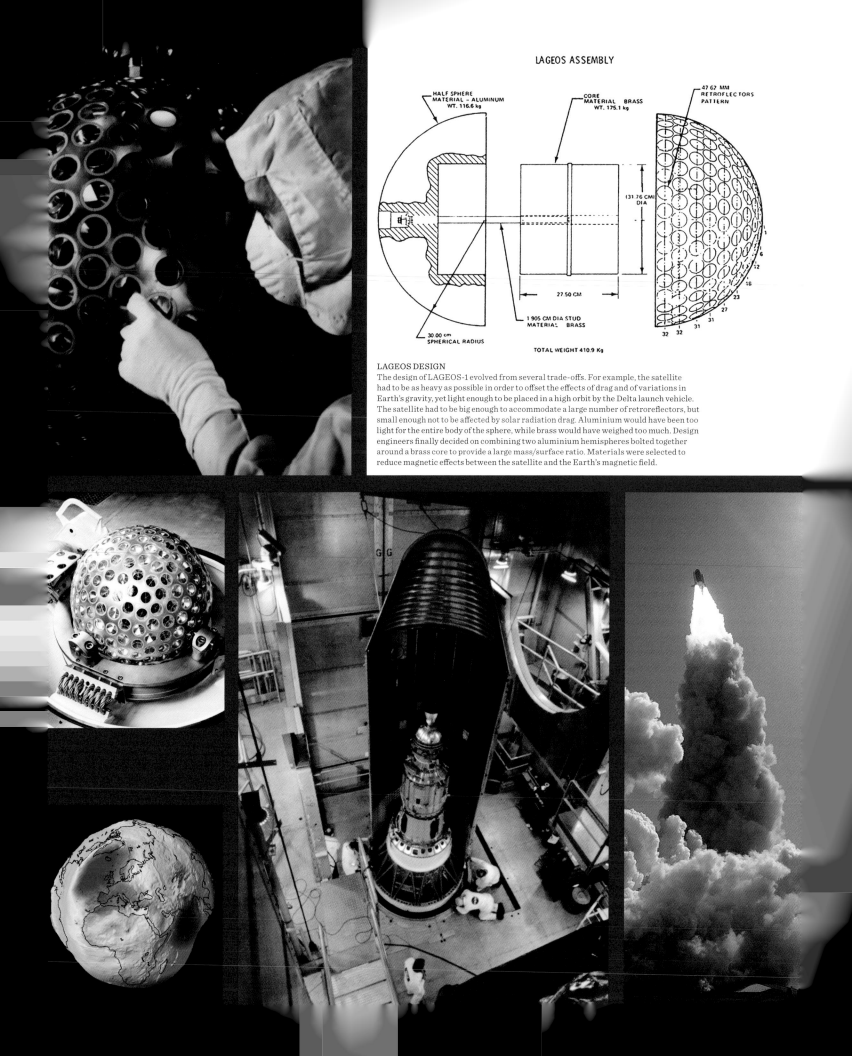

LAGEOS ASSEMBLY

HALF SPHERE
MATERIAL – ALUMINUM
WT. 116.6 kg

CORE
MATERIAL BRASS
WT. 175.1 kg

47.62 MM
RETROFLECTORS
PATTERN

(31.76 CM)
DIA

27.50 CM

1.905 CM DIA STUD
MATERIAL BRASS

30.00 cm
SPHERICAL RADIUS

TOTAL WEIGHT 410.9 Kg

LAGEOS DESIGN

The design of LAGEOS-1 evolved from several trade-offs. For example, the satellite
had to be as heavy as possible in order to offset the effects of drag and of variations in
Earth's gravity, yet light enough to be placed in a high orbit by the Delta launch vehicle.
The satellite had to be big enough to accommodate a large number of retroreflectors, but
small enough not to be affected by solar radiation drag. Aluminium would have been too
light for the entire body of the sphere, while brass would have weighed too much. Design
engineers finally decided on combining two aluminium hemispheres bolted together
around a brass core to provide a large mass/surface ratio. Materials were selected to
reduce magnetic effects between the satellite and the Earth's magnetic field.

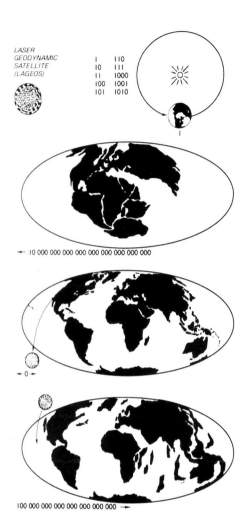

Opposite page: NASA's Laser Geodynamic Satellites (LAGEOS-1 and 2, launched in 1976 and 1992 respectively) are collecting data on the Earth's shape, orbit and tectonic plate movement. Clockwise from middle left: LAGEOS-1, a scientist working on one of the sphere's reflectors, a diagram of the satellite's construction, the launch of LAGEOS-2, the encapsulation of LAGEOS-1 and a map showing the variation of the Earth's gravity field. Colours in the map represent the differences between the theoretical value for the gravity at the surface, if the Earth was a perfectly smooth ellipsoid, and the actual measured value. They range from purple (low gravity), through blue, green, yellow, orange, red and magenta (high gravity). This model was created with data acquired by LAGEOS.

Above: The plaque included in LAGEOS-1, designed by Carl Sagan.

In 1976 NASA launched LAGEOS-1 (the LAser GEOdynamic Satellite) on a Delta rocket from the Vandenberg Space Force base in California. Normally during a satellite launch the team look on, wondering if the extreme forces of space flight will damage the fragile electronics contained within the satellite. Just a few minutes of flight can jeopardise years of work if one small component is damaged. But LAGEOS was a very different bit of technology. At first glance it almost looks like a piece of art or jewellery, a 60cm-diameter brass sphere covered in aluminium which has to have 426 cut stones embedded in its surface, giving it the appearance of a bejewelled golf ball. But inside this glittering sphere there is no complex instrumentation or electronics, no sensors or positioning technology. The only thing actually inside LAGEOS is a small plaque designed by the astronomer and author Carl Sagan that carries a variety of basic scientific and mathematical information intended to act as a time capsule, to deliver this knowledge-filled message to any inhabitants of the Earth who find it around 8.4 million years in the future. Why then? Because this is when the sphere is predicted to re-enter the Earth's atmosphere and crash down onto our planet's surface. So what is the function of this strange object orbiting almost 6,000 kilometres above our heads? Well, a clue is in the main three images on that message to Earth's future children that is hidden inside it. These are images of the continents on Earth past, present and future, with the future-facing image describing the estimated arrangement of the continents in 8.4 million years' time (so hopefully someone will be around to check if we got it right!). It's this movement of the continents through the ages that the sphere was designed to help track.

To do this, the LAGEOS satellite relies on a network of ground-based stations scattered across the globe that use laser beams to transmit pulses of light to the satellite, where the 426 reflectors then send these pulses back to Earth. The returning signals are picked up by an extensive system of receiving stations across Europe and Australia, the Americas and Asia. It's this system of send and receive that allows the time taken for the laser signal to travel to the satellite and back again to be measured with unparalleled precision, which in turn allows the distances between the ground stations to be determined with extraordinary accuracy. Mapping fixed points on the Earth with this degree of precision allowed scientists to finally confirm directly what they had long theorised: the continents are moving, day in and day out, and nothing on Earth is staying still.

The LAGEOS data has given us the ability to produce precise models of the movement of the Earth's crust that have also allowed us to peer into the future of our planet, revealing how the map of the world will appear to any generations still alive on Earth in the distant future. As well as this future gazing, the LAGEOS data has also allowed us to look backwards. By combining a whole range of data sets, including fault-line analysis, movements of the ocean floor and geomagnetic signatures from rock samples, an astonishing model has been constructed that for the first time allows us to accurately rewind the clock, enabling us to look back into Earth's past, to see how the continents have evolved over hundreds of millions of years. What this model tells us is that within a billion or so years after the onset of plate tectonics, the surface of planet Earth had been transformed.

Now this may all seem like a long way away from our story of plants, but the creation of the tectonic plates – be it by meteorite strike or some other process – was instrumental in the history of life on Earth. This is because it was the creation of granite driven by the fractured crust of the young Earth that in turn created the dynamic to transform a water world into a planet of land and sea, a planet with endless opportunities for life to now conquer.

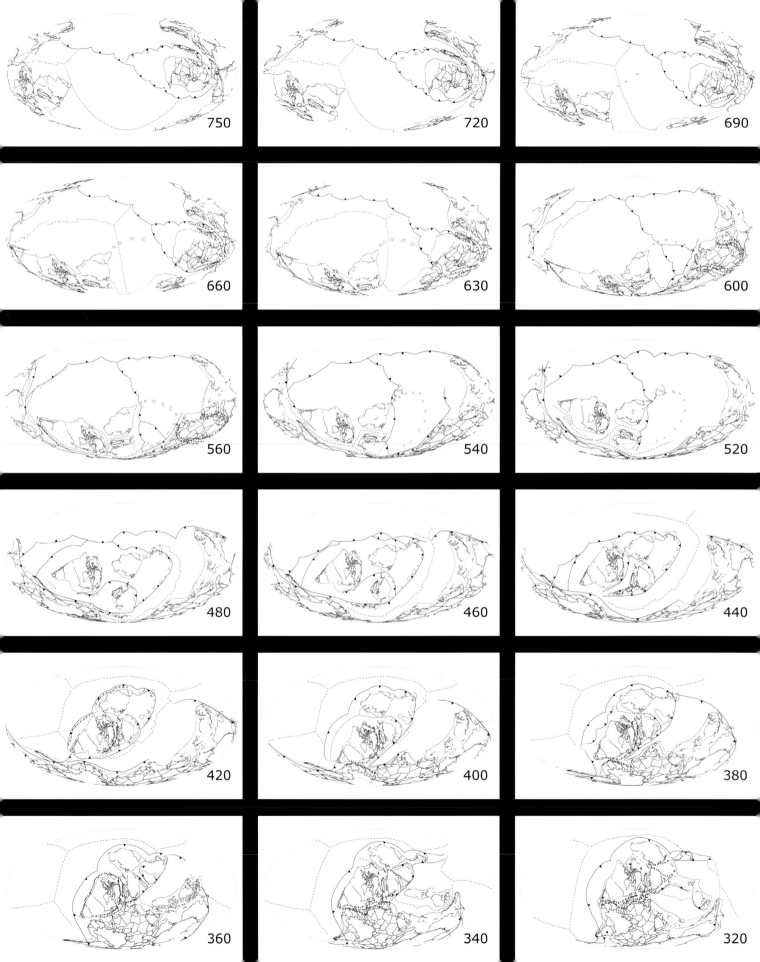

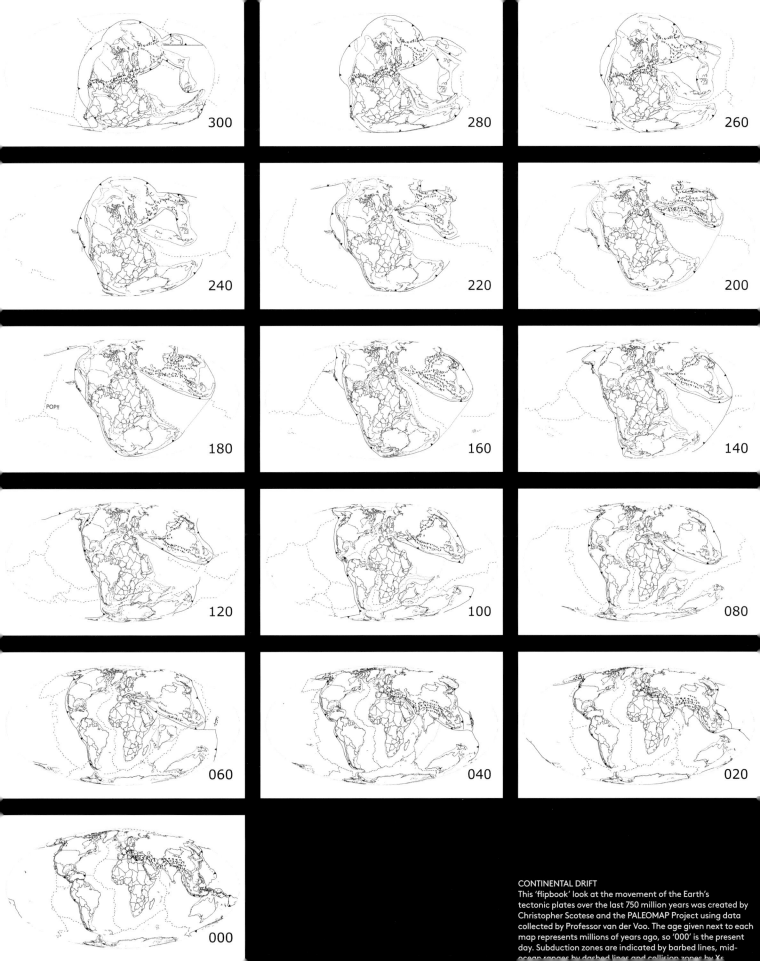

300

280

260

240

220

200

POP!!

180

160

140

120

100

080

060

040

020

000

CONTINENTAL DRIFT
This 'flipbook' look at the movement of the Earth's
tectonic plates over the last 750 million years was created by
Christopher Scotese and the PALEOMAP Project using data
collected by Professor van der Voo. The age given next to each
map represents millions of years ago, so '000' is the present
day. Subduction zones are indicated by barbed lines, mid-
ocean ranges by dashed lines and collision zones by Xs.

INVASION

'We may never know the combination of factors that spurred green algae onto land. Maybe it was the changing of Earth's landscapes to more clement conditions or the time needed for profound genetic change, but whatever the reason, green algae began to come ashore. in a journey that is likely to have started at the pebbly shorelines of freshwater lakes and rivers.'
Dr Susannah Lydon, University of Nottingham

THE EVOLUTION OF PLANT GROUPS

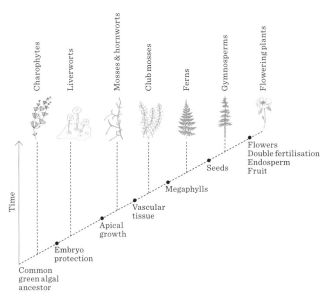

Charophytes
Liverworts
Mosses & hornworts
Club mosses
Ferns
Gymnosperms
Flowering plants

Time

Flowers
Double fertilisation
Endosperm
Fruit

Seeds

Megaphylls

Vascular tissue

Apical growth

Embryo protection

Common green algal ancestor

Opposite: *Algae and Seaweed,* illustrated in 1923 by Adolphe Philippe Millot, a French naturalist illustrator.

Fast-forward a couple of billion years from the inception of the theory of plate tectonics and the upshot of all those chaotic events was a planet transformed. One billion years ago, this was a world with a multitude of stable, continent-sized landmasses strewn across its surface, primed and ready for invasion, a new frontier packed with untold stores of nutrients within its rocky granite surface. But for all of its apparent potential, this was still a destination far out of reach, because the ancestors of every plant alive today were faced with a series of colossal challenges to make it out of the water and onto that dry land.

Just as every living thing on Earth has a universal ancestor (known as LUCA) – be it plant, animal, bacterium or fungus – so three billion years ago as the branches of life continued to divide on the planet there was one type of plant life living in the ancient oceans that would go on to be the forerunner of every plant you can see on dry land today. However, we don't know for certain what the precise universal ancestor of terrestrial plant life on Earth actually was. This is still an area of intense research – painstaking detective work that requires slowly tracing back through hundreds of millions of years of history by looking for clues in the genomes of plant life that exists today. But what we do know is that every terrestrial plant currently living evolved from a type of humble unicellular alga, an unassuming organism that would go on to radically alter the destiny of planet Earth.

We think that this ancient alga is closely related to a group of freshwater green algae that are alive today and are known as the charophytes: in particular, a class of green algae known as the *Zygnematophyceae*, the study of which organism has given us a window into the adaptations that plant life required to make it onto dry land. And we think that around a billion years ago it was plant life like this that made the first step to lifting itself out of the water and onto the land.

For millions of years before this, plant life had been happily living while locked in the oceans. As we've already discussed, life capable of photosynthesis began thriving in the oceans around three billion years ago, with marine algae that had evolved to harvest energy directly from the Sun and so accessed an almost limitless source of energy. With a distinctive green hue from the photosynthetic chlorophyll contained inside specialised cells, these highly specialised life forms not only changed the oceans but also the atmosphere of the whole planet, slowly raising oxygen levels, with all the consequences that would bring. But for around two billion years, despite their success and abundance, for these ocean-dwelling algae there was no route out of the water, so that is exactly where they would stay. It is a period of Earth's history that is often referred to as the boring billions.

Then around a billion years ago something began to change. At first it wasn't a giant leap, because these life forms were far from ready to simply step straight out of the water and take root. Instead, they took a smaller step – leaving the salty oceans behind for the freshwater rivers and lakes that had formed on those early landmasses. Here was a new habitat that the green algae could exploit, bringing them into close proximity with the land without exposing them to its full hostility. Once again life had, it seemed, reached an unbreachable boundary. For those early green pioneers, forsaking the security of water was essentially impossible, so once again, for the next half a billion years nothing happened.

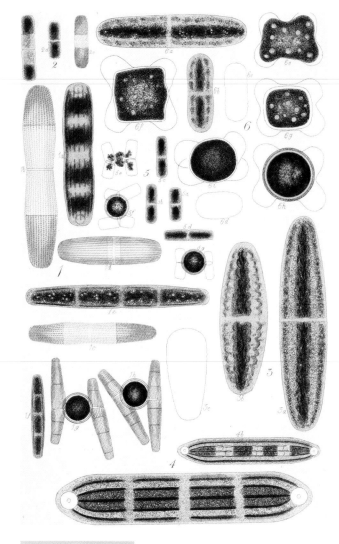

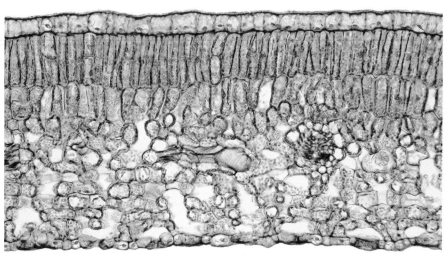

Top left: Green algae illustrated in *The British Desmidiaceae* by John Ralfs, 1848.

Top right: Cross section of a Camellia leaf, showing cuticle, palisade parenchyma, spongy parenchyma, vascular bundles and epidermis.

Left: *Penium margaritaceum*, from the work of Dr Jocelyn Rose at Cornell University.

To understand what was needed to lift these algae out of the water and onto the land, let's imagine sitting on the edge of one of those ancient landscapes around 500 million years ago. In front of you the vista is barren and rocky, a dusty terrain only broken by a large freshwater lake fed by a collection of streams running off the surrounding hilltops. In the water a vast microscopic colony of green algae thrive just as they have done for hundreds of millions of years, but it's around the edges of the lake that something interesting is happening. This is the new frontier, a boundary between wet and dry that offers a host of new opportunities but also many new challenges. To exist here plant life will need to deal with a multitude of new factors – drought, gravity (without the buoyancy of water) and the baking Sun raining deadly UV radiation down on them, to name just a few.

It was a seemingly impossible set of hurdles, but evolution would find an answer. And to understand how these immense challenges were overcome we can look for clues in the modern-day characteristics of the ancestors of those primitive algae living in that ancient lake – ancestors like *Penium margaritaceum*. This single-celled alga can today be found living in the margin between freshwater and land that belongs to the charophytes, the group of algae we know has a common ancestor with the first land plants some 600 million years ago. In 2020 an international team of scientists led by Jocelyn Rose at Cornell University sequenced the genome of this simple-looking organism, a task you would think to be relatively simple compared to the extraordinary accomplishments in genetic research of the last few decades. But in fact sequencing the genome of *Penium margaritaceum* is a greater task than you might imagine; it's larger than the 32,000-gene maize genomes (one of the largest plant genetic blueprints ever mapped), with 50,000 genes, which makes the human genome look small in comparison. The reason for its vast scale is that it has a high proportion of repeat sequences that when studied have revealed its genome to be full of a massive number of gene duplications. Hidden within this vast array of genes are clues to how a single-celled alga like this began to build the characteristics needed to get started on its evolutionary trajectory towards territorialisation.

To survive in this environment through the frequent periods of dryness in the shallow wetlands the *Penium margaritaceum* cells produce a thick, waxy, mucosal

Above: The geneticist Barbara McClintock is famous for mapping the genetics of maize (*Zea mays*) in the 1940s and 50s. She discovered the movement of genes in chromosomes by observing patterns of kernel coloration.

Right: Artichokes are particularly high in flavonoids, which protect the plants from UV damage.

coating that is vital for water retention through the dry periods. They also produce complex flavonoid compounds, one of the major ways by which land plants protect themselves from the destructive effects of UV radiation. As well as this, the genome has revealed the pathways the algae use to build rigid cell walls to combat the pull of gravity on dry land as well as the protection needed to combat the other stresses of land life, like desiccation and the more intense and potentially deadly levels of light. Put this all together, and you can see in the armoury of *Penium margaritaceum* the necessary features that would have allowed a plant cell like this to finally take those first tentative steps to life out of the water.

Around 500 million years ago it was plant cells like this that began scratching out an existence on the glimmering shorelines of some of our long-lost bodies of water. But this was a precarious life, only just hanging on through the ever-shifting environment, with its endlessly changing water levels and baking sun, every season threatening to cut short their very existence. This was an organism that had adapted to the harshest of environments but was still too vulnerable to make any substantial headway onto the land. The abiotic factors – the physical rather than biological challenges that faced these ancient plant ancestors –were too strong to allow it to fully flourish. Trapped on the edges, dependent on a constant source of water and unable to gain a substantial grip on the bare rock, the undertaking looked at best limited and at worst doomed unless this advanced guard could find something else to help them make it. And it turned out that the help they needed was to come from another branch of life that had already succeeded in carving out a niche on the exact rock where plant life was struggling to gain a foothold.

PIONEERS

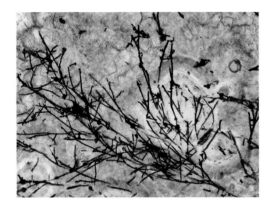

The fossil record has enabled us to trace life on Earth directly back almost three and a half billion years. Hidden away in ancient rock, the echoes of past life have emerged from the dust to reveal the form and function of organisms that have long since disappeared, allowing us to see how evolution's hand has shaped the direction of all life on Earth. The oldest of all the millions of fossils that we have discovered is of cyanobacteria, the photosynthesising microorganisms that lived in the primordial oceans of Earth just a billion or so years after our planet was formed and played such a critical role in its transformation from sterile Earth to abundant Eden.

Starting with this ancient remnant, if we could lay out every fossil we've ever discovered so that you could wander from the distant past through the ever-expanding tree of life, then you would soon notice a particular fact from the vast array of fossils laid out in front of you. For the first three billion years every fossil comes from an organism that lived out its life in the ocean; to find the first record of a land-living organism you would have to walk through these three billion years of evolution until you finally hit dry land, but that first land-living organism wouldn't necessarily be the life form you might be expecting.

In 2021 a group of scientists from Virginia Tech, the Chinese Academy of Sciences and the University of Cincinnati discovered a microscopic fossil in a sedimentary rock sample they had extracted from a cave in the Doushantuo Formation in southwestern China. In an area renowned for its abundance of fossils, this was at first glance not the most surprising of finds. But this was no ordinary fossil. It was a microfossil found in tiny cavities within rock that had been extracted from the base of these formations, which meant it could be dated to around 635 million years ago. Too small to see with the naked eye, this serendipitous discovery of tiny tendrils just one-tenth of the width of a human hair was evidence of life living and perhaps even thriving on land long before any plants or animals had made it.

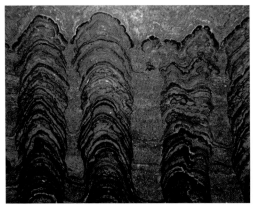

Top: Lichen, a composite organism made of algae or cyanobacteria lives among a fungus filaments in a mutually beneficial relationship – the fungus benefits from the carbohydrates produced by the algae or cyanobacteria.

Above: This cross section of a stromatolite fossil 2.4 billion years old shows a concentric banding pattern of numerous layers of cyanobacteria, which have been preserved due to their ability to secrete calcium carbonate.

Although not entirely conclusive, the evidence and analysis point to the fact that these are the fossilised remains of land-dwelling fungus that was living at a time when the planet was recovering from the Snowball Earth cataclysm. With the oceans still frozen and any exposed land providing the harshest of environments for life, it seems that somehow these microscopic cave dwellers got a foothold and began to lay the foundations for life to flourish in a terrestrial environment.

'It was an accidental discovery,' reflects Tian Gan, one of the team from Virginia Tech. 'At that moment we realised that this could be the fossil that scientists have been looking for for a long time. If our interpretation is correct, it will be helpful for understanding the paleoclimate change and early life evolution.'

This discovery provides a tantalising glimpse of the earliest pioneers of life on land, and it's not the only one. Almost all the earliest fossils of terrestrial life that have been discovered are forms of fungi, and we now think that perhaps as long as half a billion years before plants successfully made it onto dry land, fungi like *Tortotubus protuberans* were surviving and thriving on these hostile early landmasses.

So why is it that fungi endured and successfully found their way onto dry land while other life forms fell short? The answer is in the ingenious way by which fungi access the nutrient resources they need in order to live. Unlike plant life, fungi can directly access nutrients from the bare rock around them, using enzymes excreted from their microscopic filaments to break down the rock. It was this unique ability that we think enabled them to gain a foothold on the barren surfaces of the young Earth hundreds of millions of years before any other form of complex life. How long they lived their solitary life on the land we will perhaps never know. Soft-bodied organisms like fungi rarely leave fossils, so it's unlikely we will ever be able to trace back this particular part of Earth's history in any real detail, but what we do know for certain is that those early fungal life forms laid the groundwork for the first wave of plant life to make its own journey onto dry land.

POSITIVE EFFECTS OF ARBUSCULAR
MYCORRHIZAL (AM) COLONISATION

The network of arbuscular mycorrhizal fungi (AMF) extends
beyond the depletion zone (grey), accessing a greater area of soil
for phosphate uptake. A mycorrhizal–phosphate depletion zone
will also eventually form around AM hyphae (purple). Other
nutrients that are more easily absorbed by AM roots include
nitrogen (ammonium) and zinc. Benefits from colonisation
include tolerances to many abiotic and biotic stresses through
induction of systemic acquired resistance.

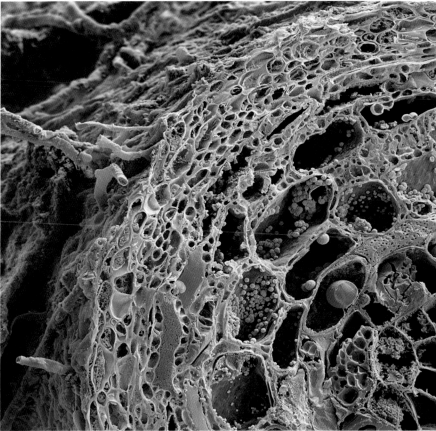

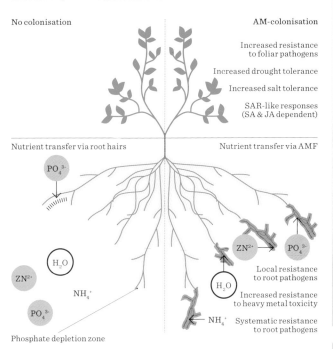

No colonisation

AM-colonisation

Increased resistance
to foliar pathogens

Increased drought tolerance

Increased salt tolerance

SAR-like responses
(SA & JA dependent)

Nutrient transfer via root hairs

Nutrient transfer via AMF

PO_4^{3-}

ZN^{2+}

PO_4^{3-}

H_2O

ZN^{2+}

H_2O

NH_4^+

Local resistance
to root pathogens

PO_4^{3-}

Increased resistance
to heavy metal toxicity

NH_4^+

Systematic resistance
to root pathogens

Phosphate depletion zone

MYCORRHIZAL NETWORK

A mycorrhizal network is an underground network created
by the hyphae of mycorrhizal fungi joining with plant roots.
This network connects individual plants and transfers water,
carbon, nitrogen and other nutrients and minerals between
participants. In natural ecosystems, 90% of plants are part of
these networks and may be dependent on fungal symbionts
for 90% of their phosphorus requirements and 80% of their
nitrogen requirements.

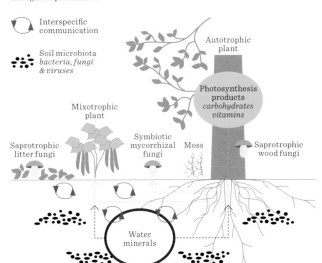

Interspecific
communication

Soil microbiota
*bacteria, fungi
& viruses*

Autotrophic
plant

Mixotrophic
plant

Photosynthesis
products
*carbohydrates
vitamins*

Saprotrophic
litter fungi

Symbiotic
mycorrhizal
fungi

Moss

Saprotrophic
wood fungi

Water
minerals

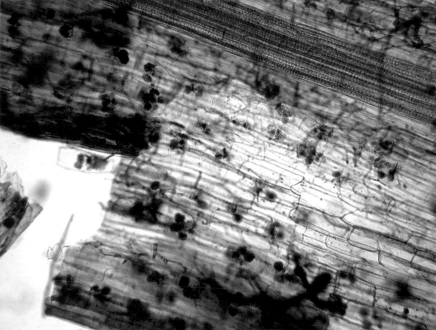

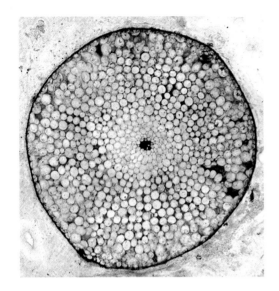

Above: Thin section of a piece of Rhynie chert, which includes a fossilised plant stem.

Opposite top: Electron micrograph image of a spruce root with symbiotic ectomycorrhiza. The soil fungus (grey) grows between the cells of the root cortex, but does not penetrate its tissues.

Opposite bottom: Arbuscular mycorrhiza is a type of mycorrhiza in which the symbiont fungus penetrates the cortical cells of the roots of a vascular plant forming unique structures called arbuscules. Shown here in the cortical cells of a flax root.

'The ancient hot springs that were here at Rhynie turned any plant material they touched into rock, which got buried over millions of years. Rhynie chert fossils are some of the only fossils on Earth where you can see this ancient link between plants and fungi.'
Professor Katie Field,
Sheffield University

To understand why fungi played such a critical role we need to understand how plant and fungal life forms have almost always been entwined. The relationship between plants and fungi is long and complex, being one that has been sustained for over 400 million years. Today we see that connection play out in the vast majority of plant families on Earth – over 95 per cent of all species of plants have a 'mycorrhizal association', a mutually beneficial, symbiotic relationship that occurs between the roots of the plant and the fungal hyphae or filaments. It's symbiotic because the plant, as the photosynthesising partner in the relationship, can provide a ready supply of high-energy organic molecules such as sugars that are spun out of its ability to harvest sunlight. Sharing these through its root system, most commonly within a structure known as an arbuscule, in return the plant gets access to an increased source of water and nutrients such as phosphorus that are not only essential for life but also far more easily accessed by the fungal hyphae than the plants' roots. It's through this exchange of key substances and the co-dependency that they create that the two have sustained their relationship for so many millions of years.

Go back 400 million years or so and we can see the evidence of this deep entwinement of fungal and plant life locked in the fossil record, which was perhaps most spectacularly illustrated in the rock samples discovered around the village of Rhynie, Aberdeenshire, in Scotland. The Rhynie chert, as it is known, is one of the most remarkable fossil beds ever discovered, offering a snapshot of Devonian life 410 million years ago, which has been preserved by the rapid immersion in silica-rich waters and hot springs that must have existed here at the time. The result is a glimpse of a whole ecosystem of plant and animal material captured within the rock. But even more than that, it has also yielded the earliest evidence of a 'mycorrhizal association', that familiar relationship between those ancient plants and fungi.

Only through this relationship was it possible for plant life to make the journey from the water's edge onto dry land; an evolutionary leap that required the development of specialised cells which allowed them to trade resources with those fungal pathfinders. And as it turned out, it was a perfect match, because long before the land was able to support a complex array of plant life through the intricate qualities of soil (more on that later), those early plant forms could live off the inhospitable hard rock with the help of those fungi.

This was a pivotal moment in the history of life on the planet; fuelled by sunlight and the rocks on which they sat, fungi and plants had come together to produce the first complex terrestrial ecosystem on Earth. The plants got from the fungi the nutrients they could extract from the rocks, and they repaid their fungal partners with glucose, the sugar product of photosynthesis, built from the energy of the Sun and carbon dioxide from the atmosphere. This symbiosis meant that plants could now survive permanently on these new landmasses. They'd finally made it out of the water and were ready to start conquering the land. But as is almost always the case with the unpredictable conductor that is natural selection, the route from those first tentative steps to world domination was far from straightforward.

A bioluminescent fungus (possibly *Omphalotus nidiformis*) glows on a tree trunk in the Australian rainforest at night. It's incredible to think that first collaboration between fungi and plants would lead to such an extraordinary relationship and one that would endure today. We mustn't think about fungi being about decay and death, as they're very much about life, particularly in forests. An extensive network of their hyphae – their roots, if you like – stretches out into the woodland, intrinsically linking with the roots of all of the trees, allowing them to share nutrients, and even to communicate with one another.

The key thing is none of these plants could survive without the fungi and the fungi themselves couldn't survive without the plants. And yet we always think of them being down here in the damp undergrowth, very much subservient to the plants towering above them. But in the earliest days of terrestrial plants, the situation couldn't have been more different.

IN THE FOOTSTEPS OF GIANTS

Below: A boletus mushroom, commonly known as a cep, or penny bun, is a tasty treat for foragers today.

Bottom: Cross section of a mushroom, *Agaricus* sp. showing its stalk and gills. The margins of the gills (black) produce the spore-bearing structures known as basidia.

How far into the past this relationship stretches we do not know, but within a few tens of millions of years of plants and fungi teaming up, something extraordinary happened to the previously barren landscape of planet Earth. To understand this phenomenon, let's return to that imaginary ancient landscape we first visited around 500 million years ago. Just 50 million years later, in the late Ordovician Period, something seemed familiar – the streams and lakes were not much changed and the green algae still endured, but the rest of the vista was remarkably different, no longer just barren and rocky. In the space of a few million years giants have begun to grow across this landscape – and thrive.

But these giants are not trees or even large plants, they were something far more alien to our twenty-first-century eyes, because in that short space of time something utterly extraordinary happened to fungi, which grew to sizes beyond our imagination.

This was a landscape strewn with huge, tree-stump-like mushrooms, an *Alice Through the Looking Glass* world, where everything we understand about the natural world was turned on its head. In front of us were the giant spikes of *Prototaxites*, a genus of fungi that dominated landscapes like this for tens of millions of years. Trunk-like structures woven from an intricate web of microscopic tubes no more than 50 micrometres in diameter, they grow up to a metre in width

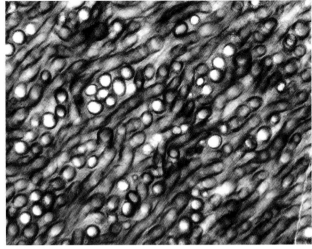

Above: Illustration of a Silurian landscape between 443 million and 416 million years ago, when vascular plants began to colonise the land. These primitive plants include bryophytes such as moss, hornworts and liverworts, *Cooksonia*, the first known plant to have an upright stalk and vascular tissue, *Baragwanathia*, a genus of extinct lycopsid plant, and the giant fungi *Prototaxites*.

Top right: Cross section of a *Prototaxite* fossil showing the tubes that make up the trunk.

Middle right: Some of the earliest land plants were liverworts like *Marchantia polymorpha*.

and over eight metres in height to become by far the largest land-dwelling organism of its time.

If you could have diverted your gaze from these giant mushrooms and looked down at the ground for a moment you would have seen that plants, by contrast, were still tiny – including small, flowerless, thick-stemmed plants with oily leaves like the liverworts, hornworts and mosses that line riverbanks today. The reason for this topsy-turvy world was simple; one crucial magic ingredient was missing from this landscape that would truly allow plant life to flourish. The bare rock played into the hands of the fungi, an organism capable of living off the rock and boosted by the presence of the photosynthesising plants. It was this life form that would grab this new opportunity and grow to newfound heights.

For the plants, however, it was a completely different story. Away from the river edges, where plants had no ready access to water, they had no hope. As soon as any moisture appeared on those impervious surfaces, it drained away into the streams, rivers and lakes, leaving the rock too exposed, too dry for plants to flourish much further. So at this point it appeared as if the Earth would be a fungal-dominated paradise for eternity. If the plants wanted to compete, if they wanted to stake their claim on the land, they would have to change the equation. They needed that one missing ingredient that we take for granted in the modern world, a magic material that we often dismiss as nothing more than dirt: soil.

MUD, GLORIOUS MUD

'If we look at the fossil record, we see something remarkable. About 470 million years ago, soils as we know them did not exist. Instead we had the earliest versions of soils, what we call proto-soils, that were only 1mm thick. They were formed from very thin layers of bacterial or fungal mats and didn't have the same building blocks that soils today have.'
Dr Aly Baumgartner, Sternberg Museum of Natural History

SOIL COMPOSITION

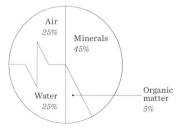

Air 25%

Minerals 45%

Water 25%

Organic matter 5%

S oil, earth, dirt; it's so ubiquitous, so abundant, so present in our lives that it's difficult to imagine a world without it. And yet soil is a complex life-support system, in fact it's the life-support system for all terrestrial plant life on Earth and therefore the basis for all complex life. Comprising of organic matter, minerals, liquid and gases, soil exists in an intricate structure of three different states: solid matter known as the soil matrix, the soil atmosphere that holds the air so crucial to the soil's function, and the soil solution that holds the precious water that all life depends on. It's a dynamic, ever-shifting system that is the product of myriad physical, chemical and biological processes that combine to create a near-magical substance that stores and purifies water, acts as a habitat for multiple organisms, helps regulate the atmosphere and, most importantly in the case of our story, acts as the medium within which plants can lay down their roots and grow.

In the modern world, there are many different types of soil, but most are made through the activities of invertebrates, fungi, bacteria and enzymes, which all work together to break down organic matter such as leaf litter or animal excrement, then mix it with minerals that have eroded from the bedrock below. The result is a substance packed full of nutrients and essentially able to hold moisture, meaning that plants can get nutrients and water all year round no matter what the variation (within reason) of the environment might be.

But 450 million years ago none of this existed; there was no soil because there were few or no animals living on land and precious little organic matter for anything to work with. And without it the very idea of plant life making further inroads was essentially impossible.

But that, as we know, is not the end of the story, because over unimaginable periods of time something began to change. Slowly but surely plants started to scratch around beneath the surface, breaking the bare rock into dust, feeding off scraps from their fungal partners and mixing that in too, then slowly filling up cracks in the rock with this new material that was being introduced to the Earth. With this came the foundations not just for the ever-evolving plant life but also a new habitat that would tempt the first few animals to make permanent homes on land as well.

But more than anything else, when it came to the production of soil, the one truly transformative ingredient was the plants themselves. Generation after generation of these early pioneering plants laid themselves down, slowly adding to the soil not just in life but in death. Decomposing where they fell, through the action of fungi and bacterial colonies, to form ever more of the magical Earth that would eventually set their descendants free.

Through this process of growth and decomposition, these patchy, thin coverings of early soils began to co-evolve with the plants and fungi that supported and created them, an Earth-based ecosystem emerging from the bare rock. It took millions upon millions of years, but eventually the day dawned when riverbanks were no longer just rocky landscapes but rich in a red-brown soil that offered plants the opportunity to evolve.

Above: A slime mould (*Physarum* sp.) plasmodium beginning to form into sporangia on a decaying beech leaf.

Above: A woodland floor with the purple fungus amethyst deceiver (*Laccaria amethystina*) growing amid the fallen seed cases and fruit of a sweet chestnut tree and leaf litter in Norfolk, UK.

Below: The root structures of vascular plants reach down into the soil to gather nutrients and to form relationships with fungi.

Below: Electron micrograph (SEM) image of a decomposing leaf (dark brown) covered with the hyphae (string-like filaments) of different fungi. There are also some bacteria (orange). Fungi are the first organisms to decompose wood because they can split and utilise lignin, a complex compound in the woody cell walls of plants.

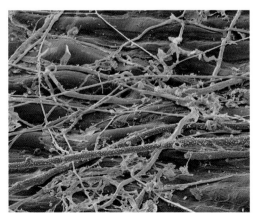

Fifty million years on and our ancient imaginary landscape has again changed; now when we look out across the river valley we see plants have a much more substantial presence. They are physically bigger, with deeper roots that utilise all of the soil's properties; to stabilise them, feed them and provide water up into the plant. Evolving over millions of years, these new vascular plant species have quite literally taken root and are now rapidly benefiting from the new niche that soil has provided. The giant fungi, Protaxites, still dominate, towering over every other living thing, but plant life is gradually gaining momentum. Some of the new plant species have also evolved to have windborne spores just like the fungi, allowing them to disperse across much wider areas, breaking free of the shackles that have chained them to the water's edge for their entire existence. And as they spread, so all of this new plant life begins transforming the landscape itself, bending and shaping the very rivers they rose from as they travel ever farther from those nursery riverbanks. All of this means that 400 million years ago, for the first time in history, significant parts of planet Earth were beginning to transform, turning into that most profound of living colours, green. But in the game of life, runaway success and sudden failure are never that far apart, and as these new green life forms hastened their colonisation of the land, there was a real danger that the newfound success would also instigate the beginning of the end for the now-verdant Earth.

You see, no matter how big we think it is, the Earth is essentially a closed system. There is a finite number of resources that get moved through different interconnected systems, which means that any significant change in one part of the system will always have a real impact on another part. This probably sounds more complicated than it actually is, because all it really means is that every action has a reaction. So, for example, if the polar ice caps begin to melt then sea levels will rise because the amount of water on Earth is pretty much fixed, it just gets moved from one form to another. What applies to water also applies to the gases in our

Below: Light micrograph of a transverse section of a horsetail (*Equisetum* sp.) leaf.

Bottom: Light micrograph of a sorus on the underside of a fern frond. A sorus is a cluster of spore-producing receptacles called sporangia (blue and green). Each sporangium is packed with spores, which are the fern's reproductive cells.

Bottom right: Illustration of a Carboniferous forest as it may have looked between 345 and 280 million years ago. The scene would have been characterised by the abundance of primitive vascular plants such as club mosses, ferns and horsetails. These reached 15 to 20 metres high and contributed largely to the formation of coal seams.

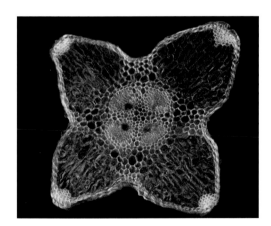

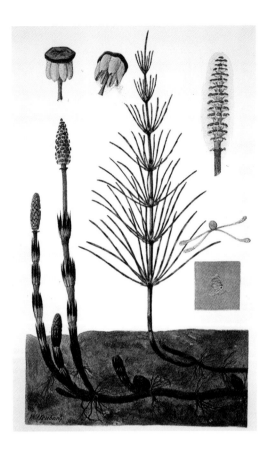

Below: A moorland in Austria overgrown with marsh horsetails (*Equisetum palustre*).

Below right: The strobilus (a sporangia-bearing structure) of a field horsetail.

atmosphere. As we know, plants use carbon dioxide to photosynthesise and produce oxygen as a waste product, so any massive increase or decrease in the number of plants wouldn't occur in isolation, it would have a profound effect on the balances of our planet – and that applies as much now as it did then. Just imagine for a minute that we were monumentally stupid enough to cut down all of the trees and poison all of the plants on Earth. The result would be that the amount of carbon dioxide in the atmosphere would rocket up along with the global temperature, and the amount of oxygen would decrease. So ultimately we would create an environment within which we wouldn't be able to breathe. Now, of course, we wouldn't be stupid enough to do that (would we?), but our species proliferation is, as we know, having a profound effect on the closed system of gases that make up our atmosphere today. We have developed new ways to exploit our success (we call it technology) and the result is an increase in atmospheric carbon dioxide that has caused an increase in global temperatures, which in turn has myriad effects on so many other elements of our climate and beyond, as we are seeing so clearly right now.

Around 300 million years ago, however, the success of plants and their rapid rise in numbers led to a very different but equally life-threatening shift in the balance of the planet. With the massive increase in the quantity of terrestrial plant life, the levels of photosynthetic activity surged and as a consequence the amount of carbon dioxide in the atmosphere began to plummet. In fact, in the first 30 million years of the Devonian Period the levels of atmospheric carbon dioxide fell by around 25 per cent.

Now, given that carbon dioxide is one of the critical resources for terrestrial plants, this had the potential to develop into a very real problem, a limiting factor on the ability of plants to grow and prosper. But as is often the case with changes to an environmental factor, the reduction in carbon dioxide also provided a new evolutionary drive, a new challenge for plants to overcome, which would lead to a period of rapid adaptation and renewal.

The collared earthstar fungus (*Geastrum michelianum*) on a forest floor in the Peak District National Park, UK. Earthstars make their appearance above ground as spherical structures, similar to puffball mushrooms. In most species, the outer layer then begins to split: at first resembling an unpeeled onion bulb. The splits become more apparent and soon the radiating arms expose an inner spore sac. The purpose of the spectacular star shape is to raise the spore sac above the surface of the ground. The collared earthstar releases its spores to the wind when raindrops strike its fruiting body.

OPEN WIDE

E volution by natural selection is a remarkably simple but also remarkably powerful concept. Small differences in the physical attributes of a living thing, caused by the endless natural process of gene mutation, go on to create a variation in the chances of the survival of that particular organism. This in turn has an impact on the chances of that organism reproducing successfully, and consequently the chances of future generations with that specific new trait reproducing successfully as well. In this way characteristics that are advantageous are conserved within a population, so gradually over vast swathes of time life forms evolve into new species – 'endless forms most beautiful', as Charles Darwin wrote in the closing pages of *On the Origin of Species.* And in this way species are born and die, and life finds a way through the cards it has been dealt.

So what did this mean for our green planet 400 million years ago? Well, as plant life surged and carbon dioxide levels dropped, the ability to harvest increased amounts of carbon dioxide more efficiently from the dwindling levels in the atmosphere would become a massively advantageous trait, which meant that plants were about to start breathing a lot more deeply.

Every plant you can see has an amazing ability to 'breathe' in and out into its environment. They do this by using stomata, tiny structures that are found all over their leaves and stems. From the Greek word for mouth, these dynamic pores are pulled open and closed by specialised cells either side of each stoma, called guard cells, that control its size. In doing so plants are able to regulate the rate of gaseous exchange with their environment by exhaling oxygen and water and sucking in carbon dioxide. When looked at using specialist timelapse photography, the opening and closing of these stomata does actually make it appear as if they really are breathing.

Around 300 million years ago, in the Mid-Devonian Period, most stomata existed on the stems of plants, but the falling atmospheric carbon dioxide level created a new dynamic that meant plants needed a greater exchange of gases with the environment in order to flourish. One potential way of doing that was to have more stomata. This was an environmental change that invited an evolutionary response, but what physical change would provide a way for plants to add more stomata to their primitive structures? The answer was as elegant as it was revolutionary: plants would respond by evolving a new flattened structure that would create a much larger surface area between plant and atmosphere, a surface that could allow a significantly greater number of stomata to open their mouths and breathe in ever-greater amounts of carbon dioxide. The answer they came up with was an extraordinary structure that we see everywhere, in the shape of the humble leaf.

The evolution of leaves provided a giant leap forward for plant life. The greater surface area meant more room for stomata to gasp in more carbon dioxide and more of it exposed to the Sun, which allowed greater levels of photosynthetic output. But with all the benefits brought about by these new structures there was also a shift in the dynamic of this leafy green landscape. With all that extra surface area a new phenomenon fell across the land, bringing with it a new level of competition. In the plant-versus-plant world of the Late Devonian, it was the shade created by all these leafy plants that transformed the contest, by bringing with it a new level of competition that fired the starting gun on a race for light, which would once again transform the surface of the Earth.

STOMATA

Stoma Open

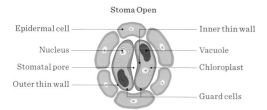

Epidermal cell — Inner thin wall
Nucleus — Vacuole
Stomatal pore — Chloroplast
Outer thin wall — Guard cells

Stoma Closed

Above: The earliest leaves of elephant's ears were nowhere near as big as this but they were a similar triumph of botanical form and function – waxy on top to keep them waterproof with a large surface area for more room for more chlorophyll to harvest more sunlight from the Sun. Underneath a strong layer kept them flat, with many stomata for better gas exchange. With these adaptations, plants thrived.

Opposite: The underside of a tropical leaf shows the clever adaptation by plants to develop large leaves to maximise photosynthetic output.

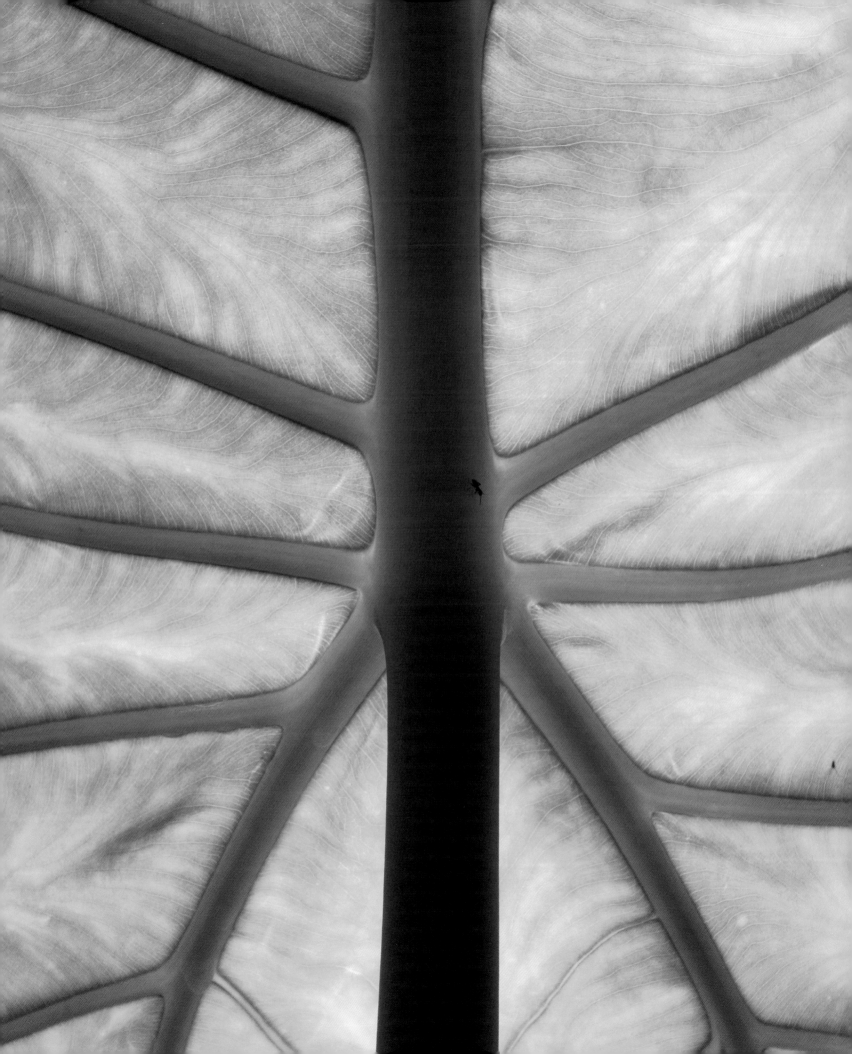

CYCLING RESOURCES

The Earth is a closed system, which means that matter is neither created nor destroyed, but constantly recycled. Water circulates through a process primarily of rain and evaporation and transpiration, and nutrients such as carbon, nitrogen and phosphorus cycle through decomposition and absorption. These nutrients are stored or 'fixed' in the bodies of plants and passed on to animals that eat them.

The Water Cycle

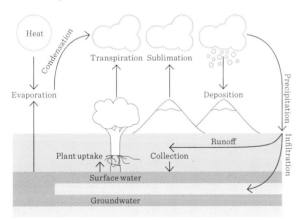

The Nutrient Cycle

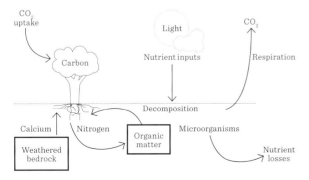

As leafy plants grew bigger and more abundant, the shade they created drove plant life to strive for one particular characteristic to keep ahead of the crowd – height. Reaching for the sky enabled plants to stay out of the shade and in the sunlight, but height requires a structural stability that doesn't come with simple fleshy stems, so it needed a sturdier substance that was becoming more and more common around 400 million years ago.

As plant structure evolved, leaves allowed photosynthesis to become ever more efficient. With more carbon dioxide coming in, more sunlight captured on those bigger surfaces and more chloroplasts to perform the photosynthetic alchemy, plants now had access to increased levels of energy that they could direct towards the construction of strong fibrous stems made of wood. This new material was becoming increasingly common 400 million years ago and would rapidly become the most effective construction material for height in the natural world.

You probably haven't heard of Cairo, a small town of just over 6,500 inhabitants tucked into the Catskill Mountains in upstate New York. Famous for very little, its most notable feature until recently has been nothing more than the oddity of its name. However, that all changed in 2019 with the discovery of one of the world's most extraordinary fossil sites, located in a sandstone quarry sitting just outside the town. Dating back 386 million years, the site was discovered by a group of scientists who came together from Cardiff University in the UK, Binghamton University in the USA and the New York State Museum. What they found preserved in the sandstone was the oldest evidence yet of a new breed of organism that emerged on the planet around 400 million years ago.

Driven by the increased levels of photosynthesis of the time, some plants had started synthesising thicker, woodier stems and roots at a far higher rate. It was this rapid growth that led to the rise of new biological mega-machines that were unlike anything the Earth had seen before; a new breed of giants whose strong wooden trunks allowed them to push their way up past the competition and towards the sunlight that every plant craves.

The oldest evidence we have of these remarkable new organisms that we now call trees is to be found just outside of Cairo. If you could have wandered through this grand ancient forest 386 million years ago, you would have found yourself in a woodland unlike anything you'd be able to find on Earth today. Surrounding you would have been a cluster of *Cladoxylopsida*, a strange-looking tree with a swollen base, short leafless branches and shallow thin roots. But this wasn't the only tree in this ancient woodland. Reaching up to 30 metres in height and 1.5 metres in diameter, another species of tree, *Archaeopteris*, was the king of the Devonian forest. At first glance this giant is reminiscent of a pine tree, but instead of the green needles we see today, the branches of the *Archaeopteris* were covered in large fern-like fronds, and at their base the trees had enormous woody roots. Stretching from New York State all the way down to Pennsylvania and beyond, this was a vast, sprawling forest, and yet this seemingly endlessly inviting habitat was unlike any on Earth today, because this was a forest virtually devoid of all animal life. It was a time when vertebrates had yet to make it onto land, there were no mammals or birds to fill the air with sound and song, so the only animal life to be found would have been primitive insects and millipede-like arthropods scrabbling around on the forest floor.

With Earth on its way to becoming a forest world, a new era had begun. The giant fungi, *Prototaxites*, that once looked down on all life on land had gone, out-competed by these all-powerful new trees on the block. From this moment, the tables permanently turned and fungi were reduced to a life in the shadows, where

Above: *Archaeopteris hibernica*, the king of the Devonian forest, had fern-like branches.

Right: The fossilised root system of *Archaeopteris* in Cairo, New York, from above (top) and with researchers for scale (bottom).

Opposite top: Cross section through the trunk of a pine (*Pinus* sp.) tree, showing the growth rings.

they have remained, working their quiet magic, ever since. This was now a world dominated by plants that had risen from humble beginnings, through millions of years of incremental gain, to become undisputed masters of the land. And as these forests grew it would mark a peak moment for green Earth. However, as we have seen again and again across the history of life on our planet, runaway success can often be a recipe for disaster; for all of their beauty and grandeur, the emergence of forests like this was something that would be far from benign.

If you've ever stood and gazed up into the high canopy of a forest, and it's something that I can thoroughly recommend, then if you're in the right place at the right time, with the right species, you might see something special. So stand, stare, blink and look for a unique pattern. You see, all of the branches and leaves from neighbouring trees don't quite meet, leaving a silvery line between them. It's almost as if they're being kind to their neighbours. It's a phenomenon called crown shyness, part of a peaceful process of evolution that has allowed all of the species in this ecosystem to come together and live harmoniously. And it works. It's beautiful.

LOCKED AWAY

Below: Illustration of a flooded Carboniferous forest containing primitive plant species like lepidodendron, an ancient lycopod also known as a scale tree. The Carboniferous forests gave rise to the coal deposits that fuel industry today.

Within the space of a few tens of millions of years, at the dawn of the Carboniferous Period about 358 million years ago, the rapid spread of terrestrial plant life had left the Earth a dramatically altered planet. The rise of trees had transformed the atmosphere, sucking vast amounts of carbon dioxide out of the air and locking all of that carbon away in the thick trunks and branches of these immense organic superstructures. As a result, atmospheric carbon dioxide levels fell sharply, triggering a period of intense global cooling and volatility in weather systems across the globe, driven by the rapidly changing climate.

This was a planet that was rapidly transforming across every one of its landmasses. In the huge southern supercontinent of Gondwana, temperatures plummeted, causing enormous glaciers to form that lowered global sea levels by as much as 100 metres. But while temperatures in the southernmost landmasses fell, in the equatorial regions the climate was still very hot and very wet. In the shadow of great mountain ranges, huge deltas had been exposed by the falling sea levels and it was in these fertile areas that vast, carbon-hungry swamp forests dominated the landscape. The forests were teeming with life and, riddled with an intense competition between species, it was here that the fight for light would lead to the

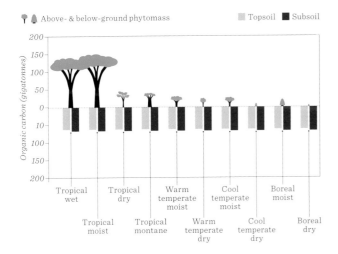

Above- & below-ground phytomass   ☐ Topsoil ■ Subsoil

Organic carbon (gigatonnes)

CARBON STORED IN ECOSYSTEMS

rise of new species of super-trees that through their success would go on to threaten the future of all terrestrial life on Earth.

Among the most successful of these new species were the lepidodendrons, also known as 'scale trees' because of the patterns found in the fossilised remains of their trunks. Reaching heights of 50 metres, with trunks up to a metre in diameter, the roots of these fir-like trees had adapted to enable them to grow both on the land and beneath the surface of the water – an adaptation that meant these giant trees could dominate huge swathes of the swamp forests.

Unlike the *Archaeopteris* trees that preceded them, the trunks of lepidodendrons were hardly made of wood at all. Instead, their interior was a soft, corky material and the exterior a robust, tough, structural shell. This was a combination that allowed them to grow incredibly quickly, reaching up to 50 metres in perhaps as little as 15 years.

But the rapid growth of the lepidodendrons, in the process sucking carbon dioxide out of the atmosphere and storing it in their mighty trunks, was just the beginning of their near-catastrophic impact. Not only did they lock away huge amounts of carbon in life, in death – as they decayed and finally toppled – the tough, indestructible shell meant that as they fell into the oxygen-depleted ooze beneath them, they didn't decompose as modern trees do, breaking down and slowly giving back their carbon into the system. Instead, the floor of the swamp forests became jammed with fallen trees and partially decayed plant matter in the form of peat to depths of more than 30 metres.

When this carbon-rich mixture was later buried under the millions of tonnes of marine sediment that was deposited by the fluctuating sea levels, all the elements were in place for a remarkable alchemy. Under intense heat and pressure, and consumed by the passage of time, this vast swathe of plant material was transformed by the Earth into a new type of rock, one that would come back to haunt us.

Above: This 42-cm-wide slab of mudstone from Carboniferous deposits near Piesberg, Germany, shows fossil branches and leaves of *Lepidodendron lycopodioides*.

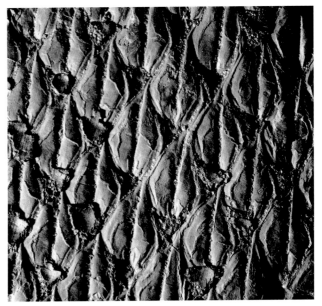

Above: This pattern, preserved in rock dating to the Lower Carboniferous Period (360 to 286 million years ago), is from the bark of the trunk of a *Lepidodendron* sp. tree. These enormous trees grew to 30–40 metres in height, but their only surviving relatives today are tiny mosses.

Above: A fossilised fern in coal shale.

Above right: A coal seam running out from the entrance of a former mine shaft at the Joggins Fossil Cliffs UNESCO World Heritage Site along the shore of the Bay of Fundy in Nova Scotia, Canada.

Today we see the evidence of these long-lost swamp forests – not on the surface of the planet but deep within it, in places like the Appalachian Basin that crosses vast expanses of the United States from New York down to north-eastern Alabama. Covering an area of around 185,000 square miles, this vast geological structure is full of the evidence of just how successful the swamp forests that thrived here were 350 million years ago. And that evidence comes out of the ground predominantly in one form only – coal. For at least the last 200 years the Appalachian Basin has been one of the largest coal-producing regions not just in the USA but the world. The mining of Appalachian coal has powered the growth of cities and of industry along the eastern seaboard and beyond, producing a concentrated form of energy created from all those millions upon millions of photosynthesising plants that lived and died here 350 million years ago. The scale of this and all the other coal deposits across the globe reveals just how vast this process was.

Throughout the 60 million years of the Carboniferous Period plants fixed carbon in the form of coal to the tune of 100,000 million tonnes every single year. Taking an enormous amount of free carbon out of the carbon cycle, these plants in life had already had a dramatic impact on the atmosphere, but in death they collectively reduced the carbon dioxide concentration from 4,000 parts per million down to just 400 per million, generating a deadly downward spiral.

Around 300 million years ago, as the Carboniferous Period drew to a close, the impact of all these carbon-hoarding equatorial swamp forests was so severe that it had left Earth teetering on the brink of extinction. Atmospheric carbon dioxide had fallen to below 100 parts per million and the planet could no longer hold onto its heat, resulting in freezing temperatures across the southern hemisphere. Around a quarter of the world's total landmasses were covered by glaciers, so the Earth sat once again within a hair's breadth of descending into another snowball event, where the reflection of the Sun's rays off the ice-covered surface could lead to the total glaciation of the planet and the extinction of all complex life on Earth.

The 200-million-year saga of plant life's adventure on land had reached its inevitable conclusion, and Earth appeared to be doomed to a lifeless, frozen fate. But just as it seemed all was lost, the planet itself would come to the rescue with the most timely of interventions, because beneath the frozen surface the giant tectonic plates that had set all these events in motion in the first place had been continuing their perpetual dance. And as they did so, they once again shifted the continents, moving most of the low-lying deltas where the swamp forests thrived away from the hot wet equatorial region and to the more temperate latitudes. In a geological instant the swamp forests began to dry up, the coal-producing conditions changed and most of these regions went out of business. As a result, atmospheric carbon dioxide began to rebound, temperatures stabilised and then rose, and the southern glaciers started to melt and disappear. The planet has not flirted with a snowball catastrophe of this scale since.

In the aftermath of all of these tumultuous events came a new world order, a more stable period of harmony between the forces that shape the planet and all life on it. Plant life continued to diversify – developing first seeds, then flowers and fruit – and in doing so forged new partnerships with the animal kingdom, transforming their signature green into every vivid colour of the rainbow and evolving specific varieties to exploit every niche on the planet.

And as they did this, plants finally found a balance, instinctively aligning the amount of biomass on Earth with the carbon cycle and the composition of the atmosphere. Plants had taken up the role as guardian of the world's climate, creating a buffer against sudden changes in either direction, breathing in and out as and when required, and paving the way for the world we've inherited today; this beautiful blooming miracle, this Eden.

Below: Blackchurch Rock lies on the north Devon coast of the UK, and is part of the Crackington Formation, an Upper Carboniferous formation of sandstone and dark mudstones and shale. The large folds were formed during a mountain-building event about 290 million years ago, called the Variscan Orogeny. The rocks were dated by cephalopod fossils called goniatites.

INFERNO

'Whether it's a massive eruption of lava or
human beings burning fossil fuels, the Earth
is going to be fine. It's the species that are
living on the Earth at the time that might
not make it through.'
Seth Burgess, California Volcano Observatory,
US Geological Survey

EXTINCTION

'Death and extinctions shape the evolution of life. It's this act of creative destruction, where one species go extinct, that allows for another species to actually rise.'
Dr Jessica Whiteside, University of Southampton

Three hundred million years ago it seems our green Earth was thriving, with the precious oxygen-filled atmosphere held in balance by the plant life that also provided the foundation for the ever-increasing complexity of ecosystems on land and sea. But even in this so-called Eden there was an uncomfortable fact about life on Earth that is as true today as it was 300 million years ago. As protective of all life as we should most certainly be, the truth is that the great diversity of our planet, the weird, wonderful and beautiful mix of species of plants, animals and fungi is only here because something else has died. In fact, because an enormous number of other things have died. If we were to take the sum total of every living thing on our planet today, it would add up to less than 1 per cent of those that have ever existed on Earth. We shouldn't always look at this colossal loss of life as a tragedy; we wouldn't have our wondrous world without it, because as painful as it sounds, extinction is a vital part of evolution.

We have an odd relationship with extinction; it's a concept that is rarely talked about in any way that isn't entirely negative (an understandable position for those of us who live to protect the natural world), but the truth is, life thrives off it. It's a critical part of the process that powers the evolution of new life forms on Earth, because if nothing ever went extinct there would be no new niches, no room for new species to evolve. In a very real sense extinction helped create the living world that we see around us today – including us humans – but when it comes to extinction our planet has always walked a tightrope: too little extinction and the process of evolution grinds to a halt, too much of a push in the other direction and the rates of extinction would go unchecked and the complex web of life begins to crumble.

We have already seen numerous examples of these nail-biting moments in Earth's history when circumstance has threatened to wipe out life – whether it's the asteroid that killed the dinosaurs, the freeze that gripped the planet, the onslaught of oxygen that threatened to rust it away, or the runaway success of plant species that could have choked the Earth. And today, of course, we are taking our own turn at controlling the wrecking ball, as human-induced climate change wreaks havoc on the planet.

From the evidence we are now seeing all around us, it would be easy to think that modern climate change is one of our planet's darkest hours, but the Earth has in fact seen far worse. New science is allowing us to delve deep into the past and reveal a series of events that led to a mass extinction on an unparalleled scale. Around 252 million years ago something caused our Earth's life support systems to fail, threatening the survival of almost every living thing. Imagine 90 per cent of all species suddenly dying – not just a few endangered plants or animals becoming extinct, or a handful of ecosystems disappearing, but nine out of ten life forms wiped off the face of the Earth. Imagine what that Earth would have looked like in the aftermath – shattered, broken, bereft of the beautiful complexity that we take for granted today. As unimaginable as this sounds, this is not an apocalyptic vision or a doomsday prophecy, it actually happened – it was a moment when the Earth turned on the life that it had nurtured for so long. In this chapter we will discover that story of unparalleled destruction, a story that can teach us much about the future of the world we see changing so drastically today.

Opposite page: A gallery of recently extinct or functionally extinct animals (those whose numbers will never recover). Clockwise from top left: dodo, passenger pigeon, Tasmanian tiger, Yangtze finless porpoises, Yangtze river dolphin, Iberian ibex, Western black rhino and small blue macaw.

A SYMPHONY OF LIFE

Our story begins just one million years before this cataclysmic event, when the Earth and all life on it was in apparent rude health. This was the Late Permian, 253 million years ago, when a view of our globe would have appeared exceptionally strange to our eyes. From one side this was a pure water world with no land in sight, but when it spun it would have revealed a landmass unlike anything we see today; this was a planet where almost every centimetre of land was clustered into one colossal supercontinent we know as Pangaea. As we discovered in the previous chapter, the idea of Pangaea was first hypothesised back in the early twentieth century by the German geologist Alfred Wegener, who noticed that all the continents on Earth appeared to fit together like a gigantic global jigsaw puzzle. Although his theory would not be proven until long after his death, decades of research, built on Wegener's initial theory, have allowed us a glimpse into the life and landscape of this early planet.

This was a world where individual landscapes may have been recognisable to us but the weather systems that swept across this huge continent made it a vastly different set of environments compared to anything we know today. In its interior, we think that the heart of the continent would have been an enormous desert, on a scale incomparable to anything on modern Earth. Locked behind massive mountain ranges, this seemingly boundless desert was almost totally dry, with all moisture and rainfall blocked by the mountain peaks. But this was just one example of a wide variety of terrains that defined this giant continent. We know from coal deposits found in the USA and Europe, which date back to the time of Pangaea, that at least parts of the equatorial region were home to tropical rainforests, an ancient

	sOrA	Angle
Arboreal nocturnal		
Terrestrial diurnal		
Terrestrial nocturnal		
Arboreal diurnal		
Aquatic diurnal		
Terrestrial generalist		

SNAKE EYES
This diagram compares snakes' eye size, 'Size-adjusted Orbital Area (sOrA)' with their field of vision (angle) to show how these traits differ in reptiles that hunt in different ways. Scientists use datasets from the study of reptiles and birds to infer much about the lives of dinosaurs based on data collected from their fossils.

Amazon-like jungle that would have been brimming with flora and fauna. In other parts of the continent flash floods and monsoon rains defined the environment and the animals that could thrive within it. But it was around the supercontinent's edges that we think life really blossomed. Here was a symphony of life, where coastal waters teemed with weird and wonderful creatures both utterly strange and yet eerily familiar, and lush coastal forests reverberated with the sights and sounds of a kaleidoscope of life forms.

In many ways, the Earth in the Late Permian was just like the planet we see today, with millions of species of animals and plants living together in complex, interconnected webs all nurturing and self-sustaining the whole, but in other ways it was a very alien world. A window into this lost world has been provided by the precious fossils we have discovered, revealing how life at this time evolved in very different ways to the Earth we see today. In the fossil record we see endless curious forms of Permian creature, like the Nycteroleter, a strange, one-metre-long animal that was part of a now-extinct group of proto-reptiles that were not quite amphibians nor reptilian. These odd creatures were insectivores feeding on a diet of ancient cockroaches, dragonflies and millipedes, and as you can see if you look closely at Nycteroleter fossils, they also had large eyes relative to their skull size, which suggests that they might have been nocturnal animals that had good night vision, but we also believe they had really good hearing; a set of characteristics that we think was highly unusual for animals of that time.

Opposite top: Pangaea around 300 million years ago.

Opposite bottom: Alfred Lothar Wegener (1880–1930), the German geophysicist and meteorologist who first proposed the theory of continental drift.

Left: Egypt's Black Desert is a volcanic field formed by tectonic activity from the Oligocene Epoch through to the Quaternary Period. It is the largest of several volcanic fields on the Arabian Plate, containing more than 800 basalt cones.

Right: A fossilised Nycteroleter in the Moscow Paleontological Museum.

In the interior of the massive supercontinent Pangaea, its heart is thought to have been an enormous desert, on a scale incomparable to anything on modern Earth. Locked behind massive mountain ranges at its edges, this boundless desert was almost totally dry, with any and all moisture and rainfall blocked by the mountain peaks. Death Valley in California, USA, is similarly bounded by mountains which prevent rain clouds from reaching it. (This phenomenon is called a rain shadow.) As a result, it is thought to be the hottest place on Earth.

Perhaps even more alien than Nycteroleter was a small dog-like animal called a Dvinia, a long-lost cousin of every mammal, including humans. Scuttling around in the scrub of the Late Permian, you might have found one of these omnivorous creatures searching for something to eat. The first fossils of this extinct genus of cynodonts were found in the northeastern Oblast region of Russia, and many subsequent finds have revealed that these animals grew up to 50 centimetres in length, with a distinctive skull shape and an anatomy featuring a large opening in the front of the eye socket that was separated from the orbit by a thin bar of bone (a feature known as a postorbital bar, a trait that only occurs in mammalian species today). This means that the Dvinia would have had forward-facing eyes, and so would have been at least in part a predatory creature, with wide cheeks to house the muscles that drove its powerful bite. Neither a mammal nor a reptile, exactly what it ate we do not know, but it's possible it might have used its jaws to crunch open shellfish just like some brown bears do today.

Without doubt one of the most impressive of specimens that has been discovered from this Permian Period is the magnificent skull of a super-predator belonging to the genus *Inostrancevia*. And what an animal this was! First found in the same Oblast region of Russia as the Dvinia fossils, this fearsome animal would have

Opposite top: First found in the northeastern Oblast region of Russia, fossils of *Dvinia* reveal it was a fantastic predator.

Opposite bottom: A mother brown bear and her cub look for shellfish in Alaska, USA.

Below: A fossil from the *Smilodon* genus of sabre-tooth cats, which lived in what are now the Americas around 2.5 million years ago.

grown up to 3 metres in length and would have been very fast-moving and able to terrorise the large herbivores of its time. What a fantastic beast it must have been, designed for speed, with powerful front legs that allowed it to run its prey down. At the moment of the kill, it's not difficult to see just how efficient a killing machine this would have been. With highly developed canine teeth, not unlike those of a sabre-toothed cat, this was weaponry designed for slashing at its prey, wounding and then waiting for it to bleed to death before ripping off and swallowing large chunks of it whole.

This is just a handful of examples of the other-worldly creatures that lived millions of years before even the dinosaurs. They give us a small glimpse into the vast biodiversity that existed 253 million years ago at the height of the Late Permian – oceans full of fish and shell-dwelling life forms and land rich in fungi, plants and an array of animals from insects to amphibians to the early cousins of all mammalian species. In many ways this may well have been our planet's first true Eden, a world where life had been given the time to develop, diversify and flourish into a fully formed biosphere. But it was not to last, because while life on Earth benefited from a period of prolonged stability, the planet itself was building towards a moment of violent transformation.

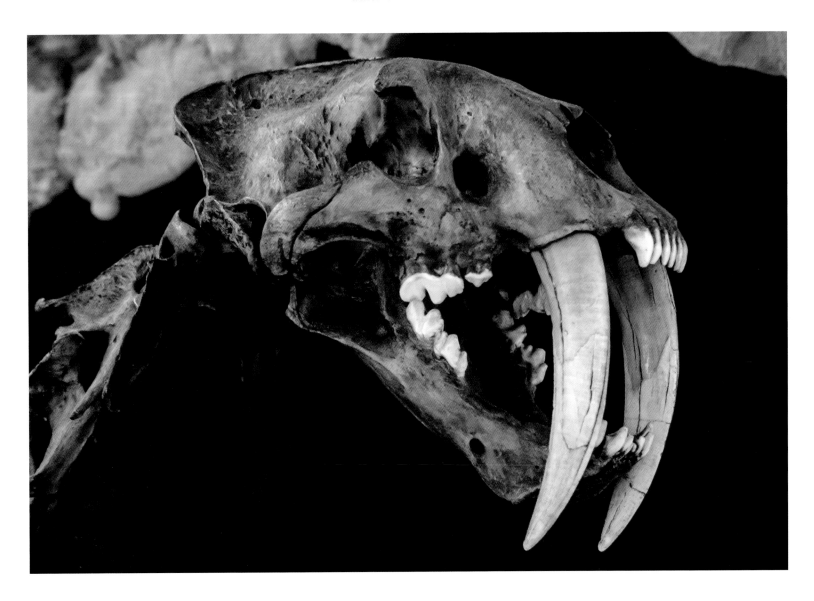

Trilobites are extinct marine arthropods that first appeared in the fossil record around 521 million years ago. They flourished throughout the lower Paleozoic before slipping into a long decline, when, during the Devonian, all trilobite orders except the Proetida died out. The last trilobites finally disappeared in the mass extinction at the end of the Permian about 252 million years ago. Trilobites were among the most successful of all early animals, existing in oceans for almost 270 million years, with over 22,000 species having been described.

AN EXPLOSIVE PLANET

Around 252 million years ago Earth was on the brink. And yet it would have been almost impossible to see any warnings from the surface, only a growing bulge in the northern hemisphere of Pangaea, in what is now modern-day Siberia, would have given the smallest indication of what was about to unfold. Exactly what event pulled the geological trigger we do not know, but what is certain is that at this time, deep within the Earth a bubble of superheated rock was slowly rising to the surface. Pushing upwards against the planet's solid outer crust, the pressure would have built up gradually until eventually the Earth's crust would have been unable to take any more and the surface of Pangaea would have suddenly and violently ripped apart. With that rupture, lava would have flooded onto the surface, forming great curtains of fire hundreds of kilometres long that would have appeared to turn the north of Pangaea inside out. These were vast volcanic eruptions dwarfing anything we have ever witnessed on the planet – even the most powerful of recent volcanic eruptions pale into insignificance when compared to those ancient events. Events like the massive eruptions we witnessed in the autumn of 2021, as the Tajogaite volcano in the Canary Islands exploded, sending tonnes of lava, ash and toxic gases into the air and shaking every centimetre of the island of La Palma. Tajogaite is a type of volcano known as monogenetic, one of a field of volcanoes that will erupt once only rather than repeatedly across the centuries. This was the first eruption on the island in 50 years, and over the course of three frightening months 170 million cubic metres of lava poured onto the surface, resulting in more than 7,000 people

Below: One year after it started erupting in 2022, the Tajogaite volcano is still degassing.

Opposite: The destruction caused by a forest fire tearing across a landscape.

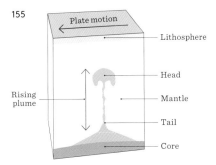

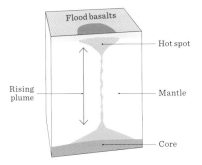

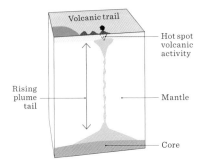

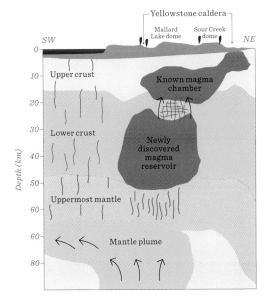

MAGMA PLUMES
The top diagram shows the process of a mantle plume's formation – this is how scientists believe the Siberian Traps could have been created. Directly above is an illustration of the magma reservoirs that currently exist below Yellowstone National Park in the USA. This explains why the region is so geologically active.

having to flee their homes. In just this single eruption the volcano spewed out enough lava to fill around 70,000 Olympic-sized swimming pools, and it covered an area of about 10 square kilometres. This all sounds impressive, but in comparison to the eruptions that were getting under way 252 million years ago in the Late Permian, this was a mere drop.

As those monstrous eruptions began in the northern reaches of the Pangaea supercontinent, around 4 million cubic kilometres of lava, ash and toxic gases were released in a series of volcanic explosions that went on for around 2 million years. This was volcanism on an unprecedented and unimaginable scale, and unlike much of our knowledge about this long-lost world, we don't have to guess at the detail of these eruptions because the evidence is still with us. Evidence that we can study, stand on and measure.

In North-western Siberia, beneath a landscape of plateaus and precipitous waterfalls, lies 7 million square kilometres of ancient lava, a massive floodplain of basalt rock that is approximately the size of Western Europe. Known as the Siberian Traps, it contains the remains of enough lava to bury the entire continent of Australia half a kilometre deep. *Trappa* is the Swedish name for a stair step; it is also the name given to the tell-tale geological feature that can be found in areas like this where repeated flows of basalt rock form a characteristic landscape of steep cliffs and narrow ledges – known as steps. Exactly what caused this scale of eruptions is still not entirely certain, particularly as it occurred so far away from the major plate boundaries, which are the normal volcanic hotspots on our planet. The leading theory is that a geological mechanism known as a mantle plume caused the release of vast amounts of lava onto the Siberian landscape. This spluttering jet of hot rock emanated from around 3,200 kilometres beneath the Earth's crust at the boundary between the core and the mantle then rose to the surface in a series of hot bubbles, carried upwards by convection through a single, long, thin channel that connected the top of the plume to its base. As it reached the surface, a bulging head formed that expanded in size as more and more of the plume rose, until it ultimately resembled a mushroom structure. This mushroom eventually met the immovable object that is the lithosphere, or crust, and started to flatten out against this seemingly impermeable obstacle, forming large amounts of basalt magma. Here it would stay, perhaps for millions of years, until eventually the calm was shattered and it broke through the surface and finally erupted over the planet.

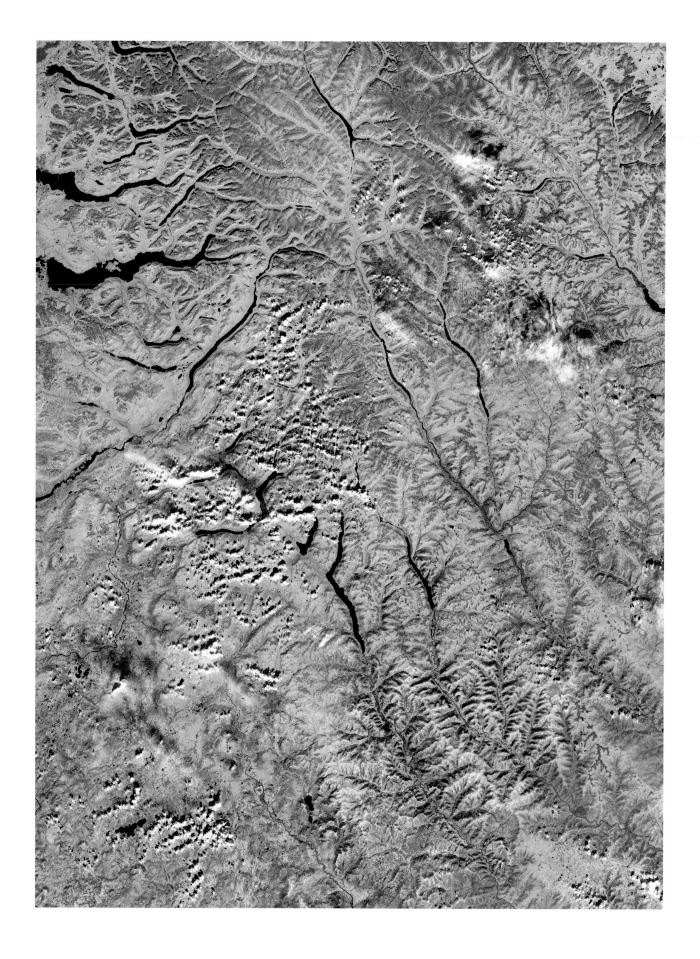

'In the 1880s Krakatoa spewed out about 2 cubic miles of magma, but even that was minuscule compared to the eruptions of the Siberian Traps – the Traps would have been like a Krakatoa erupting every year for 300,000 years.'
Dr Jessica Whiteside, University of Southampton

Above: Fires raged during the Late Permian, as volcanoes spewed magma across a dry Earth.

Opposite: An infrared satellite image of the Siberian Traps: vegetation is pink and green, depending on its water content, and water is black. This basalt landscape was formed by a massive outflowing of lava around 250 million years ago, during a massive eruption event that may have been one of the causes of the Permian–Triassic extinction.

Right: Yellowstone National Park, in the USA, is geologically very active due to the large amounts of magma stored beneath the surface. Pictured here is the geothermal geyser Old Faithful.

Above: The Harrat Khaybar volcanic field in Saudi Arabia, photographed from the International Space Station. The most recent eruptions took place between 600–700 CE. Dark regions were formed from fluid basalt lava flows, lighter-coloured areas were formed from more viscous, silica-rich lava, known as rhyolite. Sand and silt form the remaining areas.

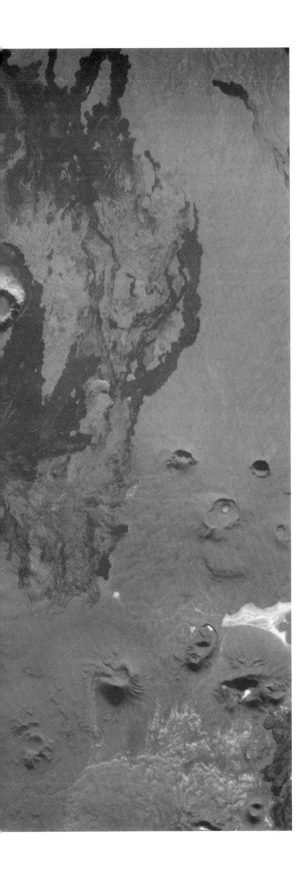

Although this is the leading theory to explain the scale of volcanic activity that created the Siberian Traps, other theories are still available! Some scientists believe it was a different trigger that sent the lava flying to the surface, one that came from outer space. Take a look at a globe and you will see that the Siberian Traps are almost exactly antipodal (on the opposite side of the planet) to a particular spot in Antarctica where we believe a crater, the Wilkes Land Crater, is buried beneath the ice sheet and is thought to have been formed around the same time as the Siberian Traps. According to the impact hypothesis, an asteroid slammed into the Earth here with such force that it was actually responsible for triggering a plume of magma to rise to the surface on the other side of the planet in what is now Siberia. Impact-induced volcanism such as this is still poorly understood, but it's an interesting alternative theory that will gain more detail as we study it further. What we can be a lot more certain about is the date of the formation of the Traps, because there is no shortage of rock samples to test from the millions of square kilometres of magma that were deposited here during the eruption. More specifically, scientists have looked once again to zircon crystals to gain the most accurate dating of the rocks, in particular using uranium within the zircon crystals as the focus of a mass spectrometry technique that allows us to date the catastrophic events that took place here around 251.9 million years ago.

It's almost impossible to imagine the impact of eruptions on this vast scale. Just locally the loss of life would have been devastating, with millions of creatures perishing not just due to the direct effects of the magma but also in the forest fires that would have ensued across thousands of kilometres of this ancient landscape. And that would have been just the beginning. As the Earth spilled out its molten innards onto the surface of the planet, clouds of toxic gas and ash billowed high into the atmosphere, not only transforming the ingredients of the atmosphere around the eruption sites but even more directly blocking out the light of the Sun. Take away the sunlight and plant life inevitably wilted and died as a volcanic winter set in, with ash falling like snow on this dying landscape, vast swathes of northern Pangaea would have been left in ruins.

Our extensive exploration of the Siberian Traps has revealed in intricate detail that these eruptions were thousands of times greater than any witnessed in human history, with an impact on the locality that would have looked like the world had all but ended. But for all the scale of the eruptions, the Earth is a pretty big place, and although utterly devastating locally, elsewhere in Pangaea we think something curious would have happened. Away from ground zero a strange haze would have been hanging in the air, a subtle indication of the nutrient-rich volcanic ash that would have been transported through the atmosphere thousands of kilometres around the globe. With this volcanic dust hanging in the air, some of the Sun's warmth would have been reflected back out into space before it could get anywhere near the surface, pushing global temperatures down by as much as 4 degrees Celsius. This combination of nutrient-rich volcanic ash and mild temperatures meant that in many places on Pangaea during the eruptions, plant life did not just survive but flourished. But the direct impact of the eruptions alone was not the end of this story – if that had been the case then this would be a chapter without anywhere else to go; the volcanoes would have stopped and life would have gradually mended itself, as it always does. But that was far from the next turn in this story.

THE LINE OF DEATH

'It's really surprising that in Italy, thousands of miles away from the eruptions in the north of Pangaea, you can still see the consequences of that event. All around the world you see this line of death in the rocks; you see how life seems to just vanish.'
M Alejandra Gómez Correa, University of Hamburg and GeoLatinas

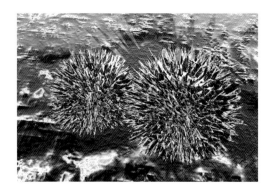

There are places on Earth of extraordinary geological interest, locations that lay out millions of years of the planet's history right in front of our eyes. One of the most exhilarating of these places is a rocky outcrop hidden away on the slopes of Mount Marmolada, the highest peak in the Dolomite range, in northeastern Italy. One hundred kilometres north of Venice, it would be easy to miss the ancient evidence that sits in front of you, just waiting to tell its story, when you look at this rock formation. These precise layers of rock formed at the same time as those ancient eruptions occurred 252 million years ago, and it's from this fossilised crime scene that we are able to uncover the next part of our chilling tale of destruction.

When these rocks were laid down in this part of Italy, they were thousands of kilometres away from the eruptions in the north of Pangaea and certainly too far away for all of that volcanic activity to have any type of dramatic impact on the life in this area. But what's interesting is that the fossil record that is frozen in time in these rock layers tells a very different story. In the lower layers laid down in the millions of years before the eruptions you can see a thick black seam of coal, indisputable evidence that life was abundant here – certainly plant life, which would have been the original Carboniferous ingredient of this fossil-fuel layer, and almost certainly animal life as well. Move upwards in time, and beyond the rich organic layer of coal every fragment contains endless microscopic fossils, the remains of plants, fungi and lots of pollen – exactly what you'd expect to find in the rich conifer forest that we think lived here. Only when you reach the geological moment in time of the eruptions are you suddenly confronted with, well, nothing.

Above all those lines of life is a grey, sandy and dull layer of rock that screams out an absence of life after so much abundance. Geologists have found this same line of death, which represents the Permian–Triassic boundary all over the planet: in China, Australia, South Africa, no matter how far away from the lava fields you travel, there's a deathly silence in the fossil records. And just down the road in another part of the Dolomites called the Bletterbach Gorge, scientists from Utrecht University have discovered another indication of the level of death that occurred here at this moment. At the same time in the fossil record, as almost all life disappeared it seems that one other form of life went into overdrive, with millions of strands of fossilised fungus suddenly appearing in the rock layers. It is thought that this sudden 'fungal spike' may have been the result of an unparalleled opportunity for the fungal scavengers of the time to feast on the abundance of decaying trees that suddenly appeared. And the team working in the Dolomites have found this spike not just here but once again all over the world, painting an even more detailed picture of the apocalyptic events that were going on 251.942 (±0.037) million years ago. Just imagine a scene where nearly all the world's trees were dying simultaneously, creating a rotting biomass of unimaginable proportions with fungi gorging on it all – and that was just the beginning. In a geological blink of an eye, almost all life in the Dolomites and across the planet was destroyed, and it would take millions of years to recover. But if it wasn't the mass expulsion of lava in the north that decimated life, what was it? The big question is what could have killed this many creatures in a geological blink of an eye, wiping out almost all life on Earth?

Above: Fossilised pollen grains are rare, but those we have found offer an insight into the early angiosperms.

Opposite top: Exploring the 'line of death' on Mount Marmolada, Italy, during filming of the series *Earth*.

Opposite bottom left: Bletterbach Gorge, formed from sandstone layers millions of years old.

Opposite bottom right: Fossils near the surface of the rocks visible in the Bletterbach Gorge.

The Marmolada is an ultra-prominent peak and the highest mountain of the Dolomites in northeastern Italy, known as the 'Queen of the Dolomites'. But its fossil record is what makes scientists flock here – notably the thick black seam of coal that proves that life was once abundant here, and also the dull, sandy grey rock layer that is the perfect illustration of the 'line of death' in the fossil record, which is an abrupt end to the presence of any life. The Permian extinction was total and very nearly absolute.

DEATH IS
IN THE AIR

For all the sound and fury of the eruptions, the biggest killer was not the molten rock bursting onto the surface of northern Pangaea 250 million years ago. The real killer was something else that was emerging from the innards of the Earth, something that had the ability to exert its influence far wider than the creeping lava fields. As those eruptions played out it was the billions of tonnes of volcanic gas being ejected high into the atmosphere that were to have the most profound effect on the planet.

All active volcanoes release large amounts of these gases, sometimes when they are dormant but most often during an eruption. There is not one source of these gases but in fact many, including those rising up from the mantle, those dissolved in the magma, those released from cavities in the volcanic rock and those generated by the heating of groundwater as the magma rises through the crust. All of this adds up to the most common components of volcanic gases being water vapour, sulphur dioxide, methane and carbon dioxide. The exact composition varies with each individual volcano, but water vapour is normally the most abundant gas, comprising around 60 per cent of total emissions, followed by carbon dioxide, which can account from anywhere between 10 and 40 per cent. As we know only too well, pumping large amounts of carbon dioxide into the atmosphere comes with serious consequences; it's an experiment we've been running ourselves for more than a hundred years. We now know for certain that the more carbon dioxide we release, the more the planet's atmosphere acts as a greenhouse, locking in heat so that the Earth becomes steadily warmer.

Around 252 million years ago the dynamic of the Earth's atmosphere was no different; more carbon dioxide in the atmosphere meant warming on a global scale. And when that carbon dioxide was being pumped out by some of the largest

Below: Along with carbon dioxide, methane is a gas associated with the greenhouse effect. The gas bubbles up from the Earth beneath Lake Baikal, Russia.

'The biggest driver of environmental change that leads to mass extinction is not the lava itself, it is the gases they release into the atmosphere.'
Seth Burgess, California Volcano Observatory, US Geological Survey

Below: Dolphins play in the Pacific Ocean off the coast of Hawaii, USA, leaping over the waves.

volcanic eruptions the Earth has ever seen, lasting for thousands upon thousands of years, the Earth's temperature began to spiral out of control. Not only that, but it's also been suggested that the levels of carbon dioxide were massively increased by the heat of the eruptions, leading to the decomposition of massive fossil-fuel deposits like coal and oil that sat in abundance in the lava fields. As if the volcanic gases weren't enough, this was the Earth consuming its own fossil fuels on an unimaginable scale, which came with an overwhelming set of complex, interlocking consequences for almost all life on Earth, consequences that we see painfully paralleled in our own climate crisis today.

To understand the deep connections that occur when a living planet starts to warm, we can choose from myriad examples all around us today, and see just how pervasive a warming climate can be to even the smartest of animals. Dolphins are one of the most adaptable and intelligent animals on Earth. Living in pods of up to a dozen individuals they can be found all across the world, thriving in a wide range of different water temperatures and environments. So at first glance you might imagine that they're not the sort of creature that would be harmed by global warming. But sadly that's just not the case. Dolphins are in trouble not directly as a consequence of the warming of the oceans due to human-induced climate change, but by the effect that the heat is having on an organism too small to see with the naked eye, on which many species of dolphin ultimately rely.

Opposite top: Blooms of the toxic algae *Karenia brevis* occur regularly along the west coast of Florida, in a red tide event.

Opposite bottom: Scanning electron micrograph image of the marine dinoflagellate (*Karenia brevis*), which causes toxic red tide. Dinoflagellates are microscopic, unicellular, often photosynthetic algae. This dinoflagellate produces a toxin called brevetoxin, which affects the central nervous system of fish.

Above: A pod of long-beaked dolphins hunt together by herding sardines into a bait ball, so they can be more easily picked off.

Dolphins are adept hunters and eat large amounts of fish, often working as a pod to herd their prey into a bait ball so that they can feed together on the captive fish. In turn, the fish they are eating have already sustained themselves on smaller fish, a food chain that ultimately leads back to its foundation of phytoplankton, otherwise known as a type of micro-alga. Most of these tiny marine organisms wouldn't hurt anything, in fact they are a critical part of the marine food chain, but there is one species, *Karenia brevis*, that can have a very different impact, because when the water warms up *Karenia brevis* goes into a reproductive frenzy, producing blooms many hundreds of times greater than it normally would.

It's a scene we're seeing more and more frequently in coastal communities, as the temperature rises and the algae flourish, painting once-blue oceans lurid greens and reds. These algal blooms are responsible for destructive phenomena like the Florida red tides that have infamously affected the Gulf Coast, stretching from Mexico to Texas and up through Florida. Extensive research into these blooms of *Karenia brevis* has shown that they are extremely toxic, producing a powerful cocktail of neurotoxins called brevetoxins that bind to sodium channels in nerve cells, causing a mass disruption of neurological processes. Any individual algae

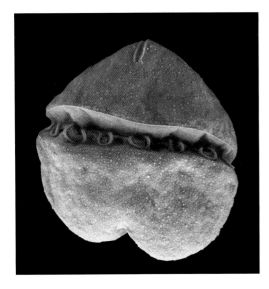

would only produce a minute amount of these toxins, but when temperatures rise and algae flourish the impact can be profound. Passing toxins up the food chain, starting with the small fish eating the algae, the toxin levels gradually build up to concentrations that in the end disrupt the nervous systems of dolphins, which can mean that in the most extreme cases the mammal can die within minutes. When it comes to global warming, it's not actually the heat that kills these creatures, it's the increases or decreases in plant or animal populations that disrupt those long-evolved stable, beautiful ecosystems. We are now seeing every day that death by global warming is not short, sharp and painless, it's prolonged and torturous. And what is true today was also true 250 million years ago.

To really understand the events that played out back then, you have to take a slightly different perspective on time. Although tens of thousands of years is an unfathomably long time for us, it is, as we have seen, in fact a very short time in the history of the Earth and the story of life upon it. And it's over these timescales that we begin to understand how the detail of that cataclysmic extinction event played out. Around 250 million years ago, 250,000 years into the extinction event, temperatures on planet Earth were slowly but surely beginning to climb from a starting point of 18 degrees Celsius (average surface temperature), rising rapidly to reach at least 25 degrees Celsius. To put that into context, the average global temperature rise due to human-induced climate change we have seen in the last 100 years is 'only' 1 degree Celsius, and yet even that small change has already wreaked havoc on the living world on land and in the oceans, so just imagine the effect of that kind of huge, rapid temperature change on life in the Permian.

Right: Ash eruption at Halema'uma'u, the summit crater of Kilauea volcano in Hawaii in 2008, the first since 1924. Kilauea is one of the world's most active volcanoes today.

Below: Marsh club moss (*Lycopodiella inundata*) in the UK (left), and Staghorn or wolf's claw club moss (*Lycopodium clavatum*) in the USA (right), are opportunistic plants that take advantage of ecosystem stress in order to flourish.

As the temperatures soared across Pangaea, the heat stress caused trees to dry out and die, leaving the once-thick canopies perforated with holes. Where there was once a dark, dank forest floor, there was now sunlight, providing an opportunity for weed-like plants such as herbaceous lycophytes to flourish. Also known as club mosses, simple but hardy plants like this thrive in times of stress to an ecosystem, when they seize the opportunity to suddenly break out of the shadows. With the forest dwindling, opportunity for new species opened up as well, with plants like the woody, seed-bearing cycads that once only grew in the tropics now blooming

'As trees die and are not replaced, animals lose precious habitat, and food supplies dwindle. There's still a lot of debate about how all the various killers came together to kill off life on land – we're talking global warming, heavy metal poisoning, acid rain, wildfires, deadly UV radiation. This was hell on Earth.'
Suresh Singh, University of Bristol

closer to the poles. As surprising as it sounds, we have good evidence that it actually made some forests *more* diverse than before the warming began, allowing a brief moment when it may have seemed that life was simply adapting to the new climate rather than perishing because of it. But this was a living world out of balance. With carbon dioxide pumping into the atmosphere in such vast amounts, temperatures continued to rise and it would only take a few more degrees of warming for ecosystems to begin to collapse.

But before that could happen we think something unexpected happened. A silence descended in the north of Pangaea, and after hundreds of thousands of years of eruptions those red-hot rivers of lava slowed at first and then turned to solid rock. Just as abruptly as they had started, the eruptions stopped.

The Permian eruptions lasted for hundreds of thousands of years, producing an inordinate amount of carbon dioxide – gigatonnes of the stuff. To put this into context, today we pump 40 billion tonnes of carbon dioxide into the atmosphere each year, a rate that scientists believe to be similar to the carbon dioxide levels released by those lava fields 250 million years ago. The consequence of our actions has pushed global temperatures up by 1.2 degrees Celsius in the space of a little over a century of fossil-fuel-burning, but back in the Permian that rate of release went on not for hundreds but hundreds of thousands of years – certainly enough to manifest global warming on a scale that would utterly devastate the planet today. An increase of 5 or 6 degrees Celsius is simply unfathomable for our planet today, but as drastic as it sounds even that temperature rise is not significant enough to account for all of the deaths that we see in the Permian fossil records. Undoubtedly there would have been mass extinctions, but not the 90 per cent of all life on Earth that we know disappeared at this time. So what happened? Well, once again, just when the Earth seemed to have done its very worst, another killer was lying in wait.

Right: Cycads are gymnosperms (naked-seeded), meaning their unfertilised seeds are open to the air to be directly fertilised by pollination, as contrasted with angiosperms, which have enclosed seeds with more complex fertilisation arrangements.

THE MORTAL BLOW

'The burning coal and the heated-up deposits of salt that were triggered by these eruptions basically created a huge time bomb.'
Sonia M. Tikoo, Stanford University

After a quarter of a million years of eruptions, some sort of calm would have returned to the battered surface of Pangaea. But beneath the desolate lava field hot magma was still flowing, forming great reservoirs underground that slowly began to bake the Earth's crust just below the surface. The inferno had moved from the exterior of the planet to deep underground, but the danger was far from over. That's because all that heat was going into rocks that were much older – hundreds of millions of years older than the land above. These rocks were rich in coal seams laid down by the life that lived here millions of years earlier, which was now igniting as it came into contact with the trapped magma, releasing ever more carbon dioxide and driving up the global temperature even further. But that was not the final tipping point of this monumental extinction event, because it wasn't just coal that was meeting the massive subterranean lakes of magma beneath the surface. There was also lots of salt, a substance that at first glance might sound far more innocuous.

Salt deposits most commonly form when ancient lakes and seas dry up, leaving behind vast seams of sodium chloride mixed up with other minerals, and when magma comes into contact with salt, things can get really nasty. When salt seams begin to bake, they release a toxic mix of halogen gases, rich in bromine and chlorine, which are not just toxic to life but also have a devastating effect on the ozone layer – one of the atmosphere's most important but fragile layers, and a critical part of the atmosphere that protects all the life below it from the Sun's most harmful ray (see Chapter 1). In effect, the magma had found its way into the worst possible place on the planet, creating a ticking time bomb that was just waiting to go off.

Trapped beneath the surface, the burning coal and the salt deposits were heated to more than 1,200 degrees Celsius. With no means of escape, this was becoming a vast, poison-filled pressure cooker, filling up every void beneath the surface until the land above could take no more. And so once again, the quiet was shattered, but this time not just by the arrival of lava to the surface but by the explosive power of those subterranean powder kegs: flinging chunks of rock 15 kilometres up into the air and filling the atmosphere with not just more climate-warming carbon dioxide but the toxic halogen gases as well, which blasted a hole in the ozone layer. And we know all too well the impact that damaging the Earth's ozone can have on life itself. In the late 1980s, our own oblivious polluting of the atmosphere created a 'hole' in the ozone layer over Antarctica that threatened to increase cases of skin cancers to pandemic levels. But this was just a hole in that protective layer. In the Permian Period the halogens released from the burning salt may have eroded away most if not all of the ozone layer, bathing every living thing beneath it in the glare of deadly ultraviolet radiation.

How do we know all of this? Well, until very recently the best evidence suggesting that the Permian eruptions blew a massive hole in the ozone layer came from a set of intriguing but indirect studies that looked at pollen grains from the time of the mass extinction. Samples of fossilised pollen taken from sites that would have been thousands of kilometres apart in the Late Permian showed remarkably similar and odd characteristics.

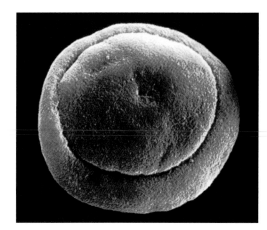

Above: Scanning electron micrograph of a fossilised pollen grain from the now-extinct conifer *Corollina* or *Classopollis* sp. Relatively abundant around 195 million years ago, these grains now define the end of the Triassic Period and the start of the Jurassic Period in the geological timescale of Earth's history. It is thought that the loss of diversity in the plant fossil record of this time, reflected in the domination of these conifers, was caused by a mass extinction event.

Above right: A glimpse into the diversity of plant pollen, from the end of the 1700s.

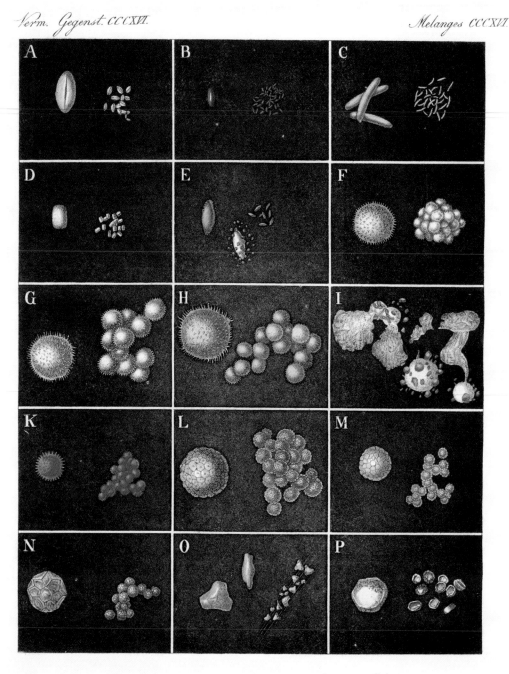

Again and again, scientists examining these ancient grains found unusually high levels of mutations in the structure of the pollen, distortions that would have significantly disrupted the function of the pollen and almost certainly have reduced the ability of the plant it came from to reproduce. If you want to induce the widespread and catastrophic collapse of a biosphere, this is a pretty efficient way of doing it. And one of the most potent ways to attack and disable the mechanism by which life reproduces is to expose it to the destructive effects of UV radiation, which penetrates cells, distorting the genetic instructions and so leading to widespread

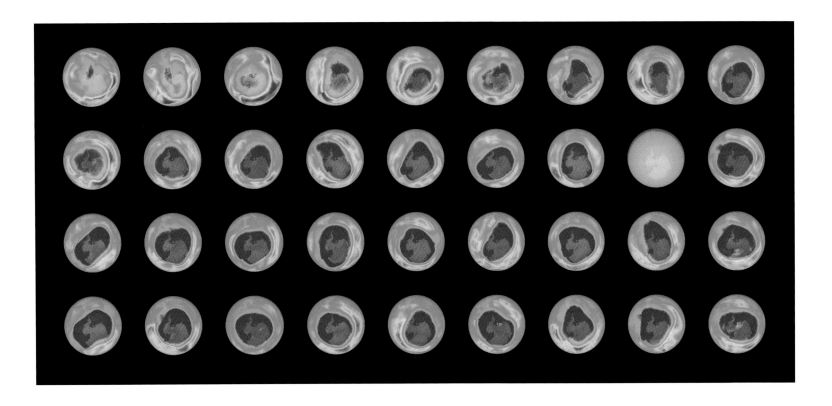

Above: Satellite images of the geographical extent of the ozone hole over Antarctica for each year (except 1995) from 1979 to 2014. Chemicals such as chlorofluorocarbons (CFCs) caused large decreases in ozone in the stratosphere. The ozone hole is now thought to have stabilised.

Below: A spiderweb photographed in the Chernobyl exclusion zone is a visual indication of radiation-induced mutation in the ecosystem.

mutation. Mutant pollen, however, only provides us with circumstantial evidence that points to the loss of the ozone layer as a major factor in the mass extinction. The release of other toxic chemicals into the environment, or factors such as acid rain or even extreme heat, could explain the increased levels of mutation that pushed life to the brink at this point. But new evidence first published in early 2023 now points more strongly than ever to ozone destruction being one of the final tipping points in this deadly cascade.

A group of scientists from China, Germany, the USA and the UK studied over 800 fossilised pollen grains from a section of rock in the Tibetan plateau that had been accurately dated to the time of the mass extinction 252 million years ago. Using a technique called infrared microspectroscopy, the team were able to delve deep into the biochemistry of these ancient pollen grains and come up with an extraordinary discovery. Looking through the fossil record, they found a significant increase in the chemicals that life produces to protect itself from UV radiation in pollen grains from the time of the extinction. This naturally occurring 'sunscreen' peaked at the precise time of the extinction, containing significantly higher levels than the pollen that was tested from earlier or later periods. This unexpected finding appears to be the first direct evidence of a major UV spike, a moment when plants began producing increased amounts of these protective chemicals in response to a significant change in the environmental conditions: a change that we are now even more certain was caused by the almost complete destruction of the ozone layer. The impact on land of this would have been utterly devastating, an unrelenting destruction of plant life that would have triggered a cascade of collapse as first herbivores and then carnivores would have starved in the face of a food chain that could no longer hold itself up. It's the type of scenario we might have faced if we hadn't acted so quickly back in the 1980s to shore up our own dwindling ozone layer – a powerful reminder that international cooperation really can change our environment for the good.

THE PERFECT STORM

So here we have it, a cascade of events that perfectly lined up to create the conditions for the most extreme extinction event in the history of our planet. First there were the eruptions – volcanic activity on an unimaginable scale that tore through the north of Pangaea, destroying almost everything in its path and transforming the landscape into the crime scene we can still see in the Siberian Traps today. Then there was the volcanic gas – less immediately destructive but with an impact that was far more widespread and enduring. The pumping of vast quantities of greenhouse gases into the atmosphere drove an abrupt and violent climatic change. As temperatures rapidly rose and the oceans acidified (as a direct result of the increased atmospheric oxygen), it was marine life that bore the brunt of the destruction at this point, in an impact that was nothing short of catastrophic. To give an example of just how overwhelming this destruction was, in one study from a sedimentary rock sample from south China it was shown that 286 out of 329 marine invertebrate genera (that's the much bigger grouping of genus, not species) were wiped out in that moment: a staggering level of biodiversity completely lost.

But that was just the beginning. While life on land had appeared to have been relatively spared compared to the devastation of marine life, the worst was yet to come. Those smouldering pyres of coal and salt released ever more carbon dioxide into the atmosphere, driving global temperatures higher and higher, while at the same time the invisible halogen gases seeped out, stripping the Earth's atmosphere of one of its most protective characteristics. At first glance the loss of the ozone layer would have appeared to change nothing, but the already fragile ecosystem was now being battered by lethal doses of UV radiation. Plant life was slowly being sterilised, with the normal cycle of life and death replaced by an endless demise. Individual plants died and could not be replaced, the biosphere was crumbling and this in turn stripped animal life of the habitat it relied on, transforming once-lush forests into ravaged wastelands. As the land fell, the oceans continued to collapse, and, as the temperature rose, the water held less oxygen, with levels dropping globally by as much as 80 per cent. Many marine creatures simply suffocated or succumbed to the ever-increasing acidity of the water. Even the life that did thrive in these intensifying conditions brought poison. In the low-oxygen oceans, it was anaerobic life that began to flourish, anaerobic bacteria that fed on the plankton blooms that appeared across the planet, in a process that filled the oceans with toxic levels of hydrogen sulphide, creating another cycle of death and decay. The once-vibrant seas became filled with rotting life forms, sinking to the ocean floor and creating a fetid bed of slime, and sulphurous tides that lapped barren shores, with a smell of rotten eggs in the air.

Below: A plankton bloom photographed from space.

Below right: Light micrograph of the dinoflagellate species *Ceratium compressum,* which causes ocean blooms as pictured left.

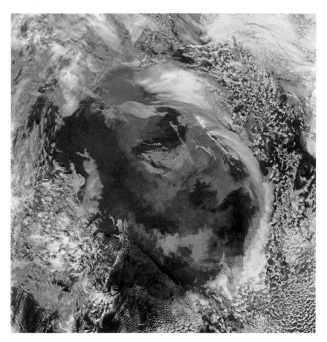

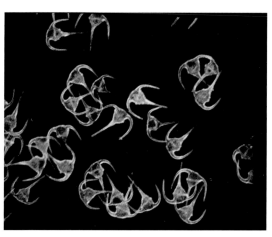

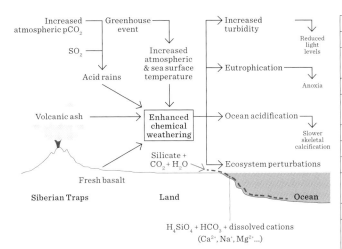

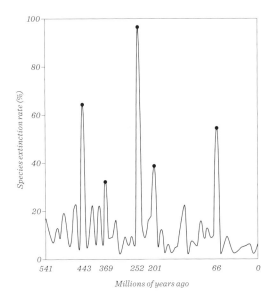

CASCADE

This illustration shows factors that contributed to the Permian extinction. Each of these factors helps to exacerbate the others, creating a 'runaway' effect.

$H_4SiO_4 + HCO_3 +$ dissolved cations $(Ca^{2+}, Na^+, Mg^{2+}...)$

SPECIES EXTINCTION

There have been five great extinction events in the history of the Earth so far. The table, top right, illustrates how complete the Permian extinction 252 million years ago was, especially in the sea.

Marine extinctions	Genera extinct	Notes
Arthropoda		
Eurypterids	100%	May have become extinct shortly before the P–Tr boundary
Ostracods	59%	
Trilobites	100%	In decline since the Devonian; only 2 genera living before the extinction
Brachiopoda		
Brachiopods	96%	Orthids and productids died out
Bryozoa		
Bryozoans	79%	Fenestrates, trepostomes and cryptostomes died out
Chordata		
Acanthodians	100%	In decline since the Devonian, with only one living family
Cnidaria		
Anthozoans	96%	Tabulate and rugose corals died out
Echinodermata		
Blastoids	100%	May have become extinct shortly before the P–Tr boundary
Crinoids	98%	Inadunates and camerates died out
Mollusca		
Ammonites	97%	Goniatites died out
Bivalves	59%	
Gastropods	98%	
Retaria		
Foraminiferans	97%	Fusulinids died out, but were almost extinct before the catastrophe
Radiolarians	99%	

Around 250 million years ago, this catastrophic extinction event was nearing its devastating completion. By the time the eruptions finally stopped for good, nearly every species had gone, and the Earth's rich complexity had diminished beyond all recognition. It's been called many things: 'The Great Dying', 'The Mother of All Mass Extinction Events' or simply 'When Life Nearly Died'. And the exact details are still a bit hazy, lost in the eons of time. Was it a single short, sharp, giant catastrophe, or a wave of smaller catastrophes? To what extent did the extinctions occur on land compared to the mass extinctions in the water? The fossil record is, of course, patchy, with much still to fill in and understand, leaving plenty of room for academic debate.

But the one thing that almost everyone agrees upon is that these ancient events caused extinctions on an unprecedented scale. This was the most profound mass extinction of all time, the moment when life almost died. Figures frequently cited suggested that less than 5 per cent of animal species survived in the sea, and no more than a third of land vertebrates made it through to the other side, all accompanied by the biggest extinction of insect life in history. At the same time, nearly all the trees on the planet perished, transforming the once-verdant landscape into a scrubland and leaving only smaller plant life to keep our green planet stumbling on. Estimates vary, but in total we think 57 per cent of all biological families and 83 per cent of all genera disappeared from the face of the planet.

But from the apocalyptic ashes there was a glimmer of hope, because for everything the Earth had thrown at it, life had survived, we know that, it must have done so, otherwise we wouldn't be here.

A WORLD REBORN

The Great Dying may have been the end of one episode in Earth's history, but it was also the beginning of another. Around 251 million years ago a new era began on planet Earth – the Triassic, a geological period that would span over 50 million years from the end of the Permian to the beginning of the Jurassic. As the curtain opened on this new Triassic age, the life that had survived came out, blinking through the dust (metaphorically speaking), and adapting to this new hostile environment and finding a way through. We don't know exactly where on the planet life endured, but we do know it was more successful towards the cooler polar regions, and on land that particularly favoured any creatures that had adapted to life underground, a place of refuge from the blistering temperatures and lethal UV radiation on the surface.

Creatures like the *Lystrosaurus*, a stocky, herbivorous animal with leathery, almost hairless skin, of which many species were no more than half a metre long (roughly the size of a small pig). It came from a now-extinct genus of creatures known as *dicynodonts*, a name that means 'two dog tooth' and refers to the *Lystrosaurus*'s short snout that contained no teeth except for two tusk-like upper canines. This was not the most gracious-looking animal; with short, sprawling limbs it would have walked more like a reptile than a mammal, but despite its ungainly manner these features made it very well suited to life underground. Those powerful forelimbs were perfect for digging tunnels, and the big barrel chest

provided lungs that could pull in large volumes of oxygen even in the dusty post-volcanic conditions. *Lystrosaurus* not only survived the great Permian Extinction but were one of the very few vertebrates to survive in its aftermath, partly down to their subterranean prowess but also, it seems, due to their ability to wander their way to new parts of the globe. Over the past 150 years palaentotologists have found huge numbers of fossils that stretch across a vast geographical area, which suggests their survival came from heading south to find new niches in which to thrive and evolve. It was a strategy that brought *Lystrosaurus* huge success in the aftermath of the eruptions, and was undoubtedly helped by the lack of any serious predation. This combination of life underground, adaptability and good fortune in avoiding being on the dinner menu made them easily the most common land vertebrate of the early Triassic. With some fossil beds found to contain almost nothing but *Lystrosaurus*, it seems that these strange animals were the main witnesses as the Earth began its long slow recovery in those dark early millennia of the Triassic.

What world did these creatures inhabit? We can only paint a partial picture of the landscape that these splay-footed creatures wandered through, but what we do know is that one plant in particular would have filled their landscape, and possibly their stomachs as well. The single group of plants that dominated the post-apocalyptic landscape was a now-extinct herbaceous genus called *Pleuromeia*. An odd and ugly weed-like plant, it had a bulbous base with short, stubby leaf-like

Left: Artwork of a scene on the shores of a lake reconstructed from fossils found in Australia dating from the Triassic. At left, lycopod trees (*Pleuromeia longicaulis*). At top right, the seed fern *Dicroidium*. At lower right, the herbivore *Lystrosaurus* (a mammal-like therapsid, widespread in the late Permian, which survived through the Triassic). At lower left, the insect *Clatrotitan andersoni* on a gymnosperm. At upper right, a female scorpionfly. At upper left, a fly closely related to the Limoniidae (small crane flies).

Opposite: The fossilised spiral 'tooth whorl' of *Helicoprion*, an extinct genus of shark-like fish that lived for around 20 million years during the Permian Period before the extinction event.

structures that produced a single cone at the top of the stem and ranged from 30 centimetres to 2 metres in length. A plant lucky enough to make it through the apocalypse and find itself with few competitors, it was the dominant vegetation in the Early Triassic, providing just enough sustenance for plant-eating creatures like the *Lystrosaurus*.

Life endured on Early Triassic Earth, but it was hardly flourishing – it was simply too hot for most life forms and in many places the planet was still uninhabitable. The fossil record suggests that ocean temperatures could have risen by as much as 14 degrees Celsius compared to temperatures before the eruptions, making the water in some places as warm as a hot bath and therefore too extreme for most marine organisms to survive. On land things weren't faring much better, either; the evidence suggests there were heatwaves driving temperatures up to 60 degrees Celsius, a heat that would have seriously inhibited the biochemical pathways of photosynthesis. Plants survived better than any other life forms through this period, but they couldn't reach anywhere near the levels that existed prior to the eruptions. In fact, in the ten million years following the mass extinction event, the fossil record contains no significant coal reserves anywhere on Earth.

Although other theories exist to explain this strange gap in the coal record, we think the most likely reason is that there simply weren't enough trees to provide the biological material required to produce fossil fuels. For life to bounce back, the planet needed one thing more than anything else: to cool down. In typical times the Earth can normally cool itself in part by removing carbon from the atmosphere through rainwater. But post-apocalypse Pangaea was mostly desert, with little rainfall, and so it took millions of years for the planet to start to cool down, and even then the world remained arid and broken, with the most vibrant areas of life largely confined to near the poles. The Earth, it seems, was stuck and needed some kind of miracle to regain the diversity of life it once knew.

Below: Wingless cockroaches climb a tree stump to survive a wildfire in Western Australia. Like *Lystrosaurus*, cockroaches are widespread across the Earth, and are consummate survivors.

Below right: This Saudi Arabian desert is a good comparison for the deserts of Pangaea in the Early Triassic, when temperatures soared and rain was scarce.

IT NEVER RAINS BUT...

The expanse of time between us and a multitude of undiscovered events stops us from being able to write the full biography of our planet, but all those gaps are part of the pleasure, as we scratch around in the dirt in an attempt to complete the story. And every now and then science is able to add a new and often unexpected chapter to this extraordinary tale. One such chapter has only recently come to light, and it concerns an event that, it is fair to say, may be one of the strangest in the Earth's long history and yet wasn't known about even just a few decades ago.

This transformative event was first discovered, almost by accident, in the 1980s when three British scientists were chatting about their work in an office in the Earth Sciences department of the University of Birmingham. Two of the scientists there that day, palaeontologist Michael Simms and geologist Alastair Ruffell, were working on two seemingly disparate areas of research. Ruffell had been looking at a set of Triassic rocks known as the Mercia Mudstone Group, a sandstone and siltstone bed of rock that was laid down between 250 and 200 million years ago and that underlies much of central and southern England. As a window into the Triassic world, Ruffell could see that the vast majority of the layers in these rocks reflected a hot, arid climate for much of that time – exactly what you would expect after the Great Dying. But in layers of rock from around 232 million years ago, during a period known as the Carnian Age, there was evidence of a very different climate. In the Carnian section of rock it seemed that the part of Pangaea that is now England was far from dry. A thin, grey layer of sandstone full of marine fossils suggested that this was not a desert but a vibrant river or delta teeming with life, 'Slap-bang in the middle of all this horrible arid stuff was this probably rather pleasant environment,' according to Ruffell. It seemed an extremely dry environment had suddenly become very wet and remained that way for over a million years.

On the other side of the room Michael Simms was listening to the picture being painted of this odd moment 232 million years ago and realised it appeared to line up with a very different event that he had been studying. Simms's research was into a group of marine animals known as crinoids, feathery-looking sea creatures that are

Opposite page: Crinoids are marine echinoderms dating back to the Lower Ordovician Period (500 million years ago). Thousands of extinct crinoid species have been found as fossils, but only a few hundred species exist today. They attach themselves to the seafloor with their stem and use their feathery arms to filter food from the water.

Right: Detail of the so-called Watchet Fault on the Somerset coast, in the UK, a striated cliff consisting of Mercia mudstones.

related to starfish, sea urchins and sea cucumbers. Crinoids still very much exist today, but Simms was interested in the lives of these creatures back in the Triassic and he'd noticed a period when there seemed to be a significant level of extinction of these creatures that occurred right in the middle of the Carnian, exactly at the time that Ruffell had found the change in climate. As Simms and Ruffell talked, it seemed that their work on events that took place 232 million years ago might not just be coincidental, but intricately linked. Intrigued, Ruffell and Simms began investigating further and found evidence for this sudden change in the environment from dry to wet in rock samples from all over the world. It appeared that it was not just wet in England 232 million years ago, but everywhere they looked: a change in global climate that was mirrored by a minor extinction event of plants and animals at exactly the same time. Ruffell and Simms called it the Carnian Pluvial Episode, a geologist's way of naming a moment in the Earth's history when it literally poured with rain.

'The skies became heavy with
moisture, the clouds burst and it
rained. A lot of rain. This episode
lasted for about 2 million years. A
deluge on this scale seemed so unlikely
that when scientists first discovered
the evidence, some believed it was just
a local event. But evidence later popped
up in China, Iran and North America,
suggesting it was a global event.'
*Evelyn Kustatscher, Museum of Nature
South Tyrol*

Above: After the Permian extinction, the Carnian
Pluvial Episode covered the Earth with water and
led to an explosion of new species, many in the sea
– and most of them dinosaurs.

It took decades for this obscure theory to turn into a major area of research, but
today it is widely accepted that 232 million years ago, in the middle of the Triassic,
something very strange happened. After millions of years of drought and desert,
it rained, pretty much everywhere, for over a million years. And now that we are
aware of this massive downpour we can see that the most overt evidence for this
massive downpour has actually been right in front of our eyes the whole time.

Wherever we look at rocks from this age there is evidence of wet weather,
and nowhere is this more dramatic nor more visible than high in the Dolomite
mountains in northern Italy. These mountains began forming very shortly after
the mass extinction event 252 million years ago, made up of layers of sedimentary
rock that sit on top of one another, each corresponding to a time in Earth's history,
with the oldest rocks at the bottom, dating to around 238 million years ago, and the
youngest at the top, which are around 200 million years old.

The initial work of Ruffell and Simms looked for evidence of a shifting climate
inside rocks like this, but here in the Dolomites you don't have to break the rocks
open to reveal their secrets; the secrets are held in plain sight and it's all to do with
the shape of the mountains. Many of the mountain tops in this part of the Italian
Alps have a characteristic shape – they start steep at the bottom, rise sharply, then

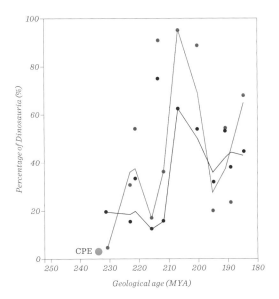

● Specimens   ● Genera

THE SUCCESS OF THE DINOSAURS
This graph plots the percentage of fossil species and distinct genera of dinosaur that have been discovered. At their peak, between 210 and 200 million years ago (MYA) the dinosaurs consistently account for more than 90% of species in the fossil record and more than 60% of the genera. The CPE refers to the Carnian Pluvial Episode, when the climate changed, temperatures and rainfall levels rising. The dinosaurs arose out of this period.

there's a characteristic shallow shelf before they rise steeply again up to the peak. It's that shallow slope that is of such interest, corresponding to about 2 million years in the Earth's history, starting around 232 million years ago.

At first glance these shallow slopes are simply evidence of the sandstone rock that's been eroded here over the years. It's a soft sedimentary rock, which when exposed to water washes off the land, down into the rivers and out into the sea. But the interesting thing is, when you date the erosion here and look at the depth of this erosion, it becomes clear that vast amounts of this sandstone were washed away. In places the erosion is 80 metres deep, which indicates one thing, a massive amount of rainfall dropping here for an incredibly long period of time. The endless rain that fell over hundreds of thousands of years has carved its unique signature into these mountains for us to read 232 million years later. We still don't know why it started. One of the leading theories suggests that it could have been underwater volcanic activity disrupting the water cycle, but at some point in the Triassic, a time as we have seen that was famed for being excruciatingly hot and arid, it started to rain. Boy did it rain! And that rain defined the Earth's climate for around 2 million years.

Below: The Dolomite mountains in Italy got their unique shape from erosion caused by endless rain over hundreds of thousands of years – beginning with the Carnian Pluvial Episode.

It was a moment that transformed the Earth for ever, a change not just to the climate but also an evolutionary kick-start that led to a dramatic shift in our living planet. Eighteen million years after the mass extinction that marked the moment when life nearly died, the heavens opened and the sudden increase in rainfall right across the long-lost supercontinent of Pangaea led to the Earth being reborn. Where arid shrubland once stood, lush forests now grew, rivers flowed and life blossomed. It's been called 'The Greening of Triassic Earth', the dawn of our modern world. There were still no flowers, grasses or birds, but in this evolutionary eruption there was a whole host of new species of plants, crocodiles and amphibians and even, we think, the early ancestors of our lineage: mammals. But it wouldn't be the mammals

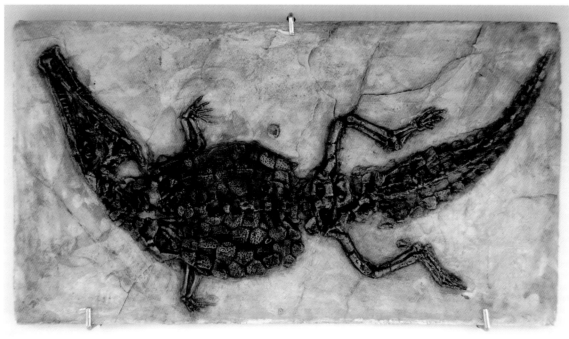

Right: The modern crocodile (top) and fossils of one of its ancestors, *Dyrosaurus phosphaticus*, which date from the Eocene (56 to 34 million years ago) and are found in North Africa.

Opposite top: Fossilised footprint of a dinosaur (Carnotaurus) at Phu Faek national forest park, Kalasin,Thailand.

Opposite: A palaeontologist inspects the femur of a Jurassic Sauropod after it was discovered during excavations at the site of Angeac-Charente, in southwestern France. The 140-million-year-old, 2m long, 500kg femur of the largest herbivorous dinosaur known to date was discovered nestled in a thick layer of clay.

who would inherit this brave new world; our time would come much later. Instead, another group of creatures was about to become the new dominant force on the planet.

It was at this moment that an apparently insignificant group of reptiles began to evolve. These reptiles were, of course, the early dinosaurs, the creatures that would ultimately lead to the tyrannosaurs, brontosaurs and stegosaurs and all the other mighty dinosaurs that would come to dominate the planet in the Jurassic and Cretaceous Periods to come. Exactly how the Carnian rains influenced or even drove the events that led to the rise of the dinosaurs is uncertain. We are not even sure if dinosaurs existed at all before the rains, and if they did it was only as an obscure group of reptiles in the south of Pangaea. Evidence indicates that these dinosaurs emerged before the Carnian, about 245 million years ago, but those earliest creatures are very rare and only a few such species are known. But in the millions of years after the rains there is no doubt that they prospered; in fact in some places 90 per cent of the vertebrate fossils we have discovered are dinosaurs. So why did they suddenly do so well? We don't know the answer to this, but it could, at least in part, be down to a change in the availability and type of food they could access, and of course the level of predation they faced. But the most dramatic and contentious claim is that the Carnian was responsible for the dinosaurs' rapid evolutionary expansion.

At the start of the Carnian, the dinosaurs were nothing more than small reptiles, and it was a different group of crocodilian reptiles called the crurotarsans that dominated this moment. But 4 million years later, by the time the rains had stopped and a new Eden was emerging, two new major groups had emerged. These were the ornithischians, a mainly herbivorous order of dinosaurs, which would go on to include iconic creatures such as stegosaurus and triceratops, and the saurischians, which gave rise to the giants, including almost all the great therapod carnivores such as *Tyrannosaurus rex* but also the huge, long-necked species such as the brachiosaurs. It was an evolutionary shift that coincided with the great rains, and after this rapid rise it was the dinosaurs that would go on to rule the planet for the next 150 million years.

EVOLUTION OF DINOSAURS
A visual representation of how the dinosaurs arose and evolved between the Triassic and the Cretaceous Periods, and how they are classified.

E early epoch   *M* middle epoch   *L* late epoch

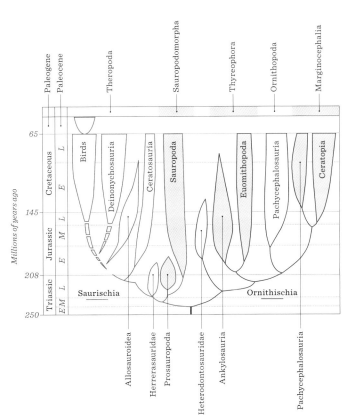

Above: Fossils are formed over thousands or even millions of years when an animal's body is buried by sediment, usually mud, sand or silt, then additional layers build up on top, putting pressure on the layers below. Soft materials such as skin, muscle and cartilage decompose and the bones and teeth turn to stone as water seeps into them from the sediment, introducing mineral elements. The sediment becomes sedimentary rock, surrounding the fossils.

Opposite top: Fossilised coral in limestone in New Caledonia (left) and the Seychelles (right). Corals first appeared in the Cambrian Period, but their fossils are extremely rare until the Ordovician Period, 100 million years later.

Opposite bottom: Artwork of how Earth's landscape might have looked in the Triassic Period, with myriad types of dinosaurs roaming the land and skies.

Dinosaurs weren't the only group to leave the Carnian age in robust evolutionary shape. In the oceans coral reefs had dramatically shifted to become much closer to the species we see alive today, and even the basis of the ocean food chain, marine plankton, had become far more 'modern' through this transformative period. And, of course, mammals, although almost hidden and insignificant at this time, were appearing in an increasing number of habitats on their long ascent to dominance. As Michael Simms, one of the first discovers of the Carnian Pluvial Episode, said, 'It was almost like a turning point between some elements of a more ancient world and a modern world.'

So what does this great period of death and destruction, followed by such profound rebirth and revolution, tell us about our planet today? Well, in one way the End Permian extinction exposed the fundamental fragility of life, as countless species of plants and animals were wiped out by those volcanic eruptions and it took millions of years for them to recover. But perhaps the event also teaches us another lesson, one about the tenacity of life, because the living world did bounce back, to be just as complex and just as beautiful. In fact, maybe even more so.

And that is where this ancient story really does begin to speak to the world we see before us today, because there's a big unspoken question that resonates all the more with the knowledge of the Earth's past that we now have. As we all know only too well, for over a hundred years we've been pumping carbon dioxide into the atmosphere and watching as global temperatures creep up, with all the consequences that we are now witnessing and experiencing. So what can those events 252 million years ago teach us about our own climate emergency?

And so we move to the final part of our planets story, a chapter where we humans finally take centre stage. It's a story that begins with the dramatic events that allowed us to inherit a world dominated by dinosaurs, shifting the balance on Earth once more into an age that would see life take the upper hand in shaping the planet's destiny.

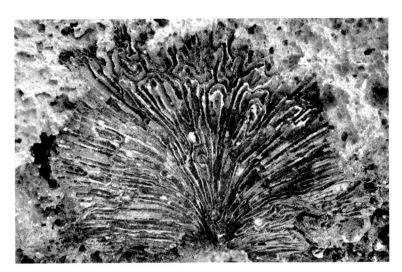

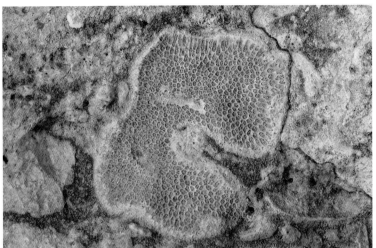

The big lesson we've learned is that whatever we do to this planet, ultimately the living world will be fine. If we don't address the climate and biodiversity crisis; if we burn every last lump of coal or drop of oil; if we turn this into a complete hellscape, life will bounce back. It will be wonderful and beautiful again. But that begs the question: should we bother? Do we need to preserve and protect this? Well, think of all of the suffering, think of the wastage. You see, that apocalypse 250 million years ago was caused by chance volcanic eruptions, part of our planetary processes. Do we really want to be the cause of an apocalypse on that scale? Do we want those sorts of extinctions on our conscience?

HUMAN

'The fate of humanity ultimately rested on
the shoulders of these animals as they braved
the new world after the asteroid collision.'
Professor Steve Brusatte,
University of Edinburgh

WINDOW ONTO THE PAST

In southwestern France, hidden deep beneath the Pyrenees mountains in the Tarascon basin, just south of the town of Foix, is an underground world known as the Cave of Niaux. Comprising more than 14 kilometres of subterranean passages and chambers, this is a labyrinthine network in which it is easy to get lost if you don't have any knowledge of the system. But to come here is not just to wonder at the intricacies of the geology in this beautiful part of France; walk through the cave system and you will eventually discover the reason why so many people want to make a pilgrimage to this special place.

After stumbling through a network of interconnecting damp and dark caves you eventually arrive in an enormous and remarkable cavern known as the 'Salon Noir' – or the Black Hall, to give it its less mysterious-sounding English name. This location was one of the last places we visited during filming, and without doubt it was one of the most beautiful and moving. The scale of the Salon Noir feels special in its own right, but it's when you shine a light onto the walls that the magic of this cave really comes alive.

'One major event can have ripple effects throughout the rest of history, and I think this event, this catastrophic impact, is almost unmatched.'
Aisha Morris, National Science Foundation

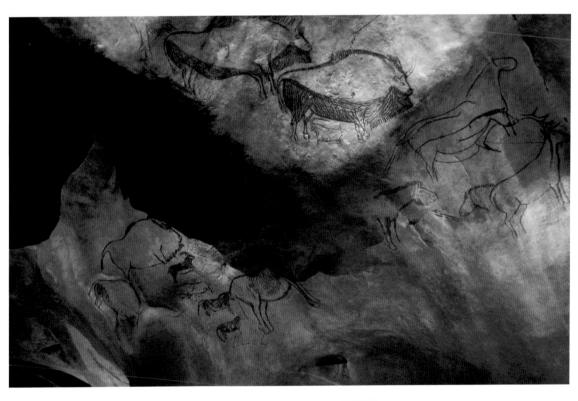

Right: The fantastic artwork found in the Cave of Niaux, in the French Pyrenees, reflects the creativity and complexity of earlier humans, as well as their esteem of the natural world.

Right: The entrance to the Cave of Niaux, which was only discovered by Western science in 1869, is located on the steep side of the Tarascon basin.

SCIENTIFIC CLASSIFICATION

Kingdom:	Animalia
Phylum:	Chordata
Class:	Mammalia
Order:	Primates
Suborder:	Haplorhini
Infraorder:	Simiiformes
Family:	Hominidae
Subfamily:	Homininae
Tribe:	Hominini
Genus:	Homo
Species:	*H. sapiens*

Binomial name

Homo sapiens

Linnaeus, 1758

Discovered in 1869 by a Commander Mollard and his two sons, the walls of the Salon Noir contain an extraordinary collection of cave art – intricate illustrations including eight beautifully depicted bison to primitive horses known as tarpan, as well as an uncannily accurate illustration of Pyrenean ibex. Beyond their simple accuracy and beauty, what makes these charcoal drawings so moving is that scientists have dated them and found that the humans that drew them in this sheltered area of the Pyrenees were living here 13,000 years ago.

When we look back at the story of the Earth, a journey of almost four and a half billion years, 13,000 years seems immaterial – a mere blink of an eye in geological time. And yet the humans that stood in this cave and the art they felt compelled to create sits at a turning point in the history of our planet and our species, a moment as important as any of the other monumental events that we have explored in this book. Whether giant asteroid impacts, vast volcanic eruptions or freezing oceans, be in no doubt that the arrival of *Homo sapiens* is right up there in terms of impact on Earth. That's because from that moment on we were on course to become one of the most powerful forces on the planet – a force for both good and bad, and one that has become capable of transforming and reshaping our Earth in ways far beyond anything that simple geology has been able to do over the last four billion years.

In this chapter we are going to explore how one species, our species, was able to rise to such power, and how it could have turned out so differently if our mammalian ancestors had continued to play second fiddle to an equally successful set of creatures – the dinosaurs. It's a story that starts with destruction, with an event of such magnitude that it decimated one world and in doing so allowed a new one to take its place. And through the endless chaos and upheaval the planet threw at this new world, we will see how Earth has blindly conspired to drive the course of evolution towards the creation of our species, an animal so powerful it would come to dominate all other life.

Right: School children around the world now learn about palaeolithic art to better understand human history.

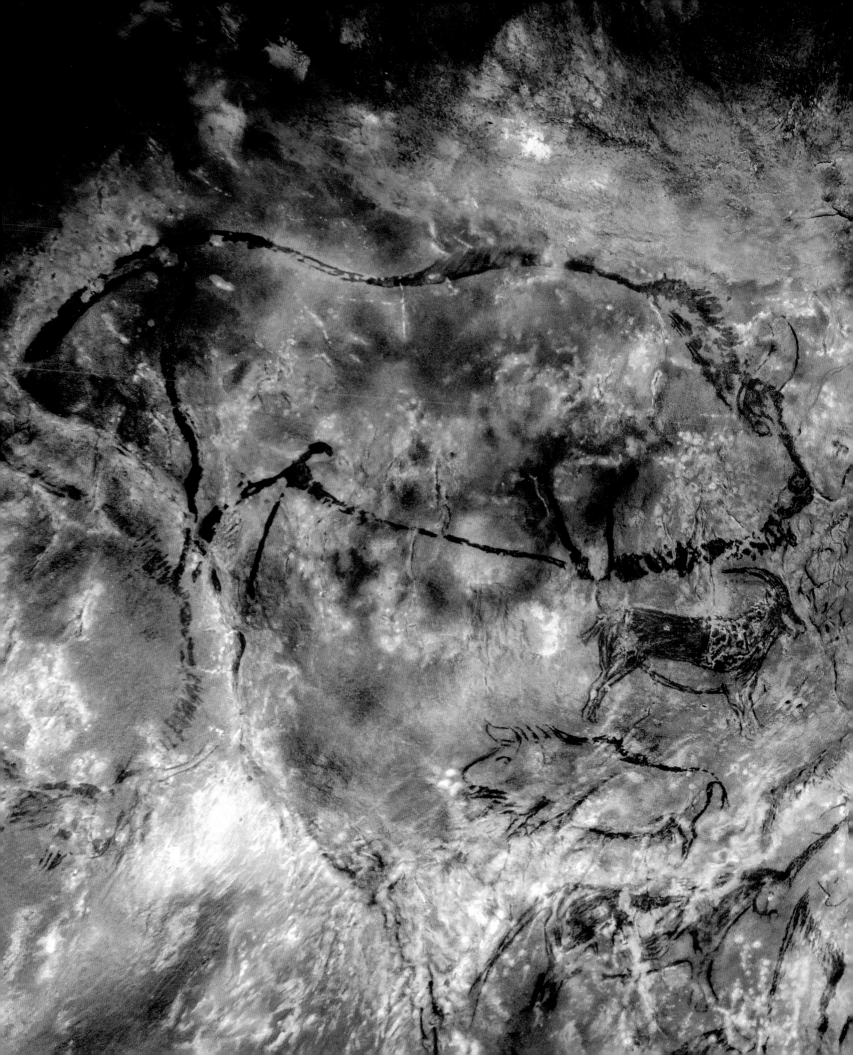

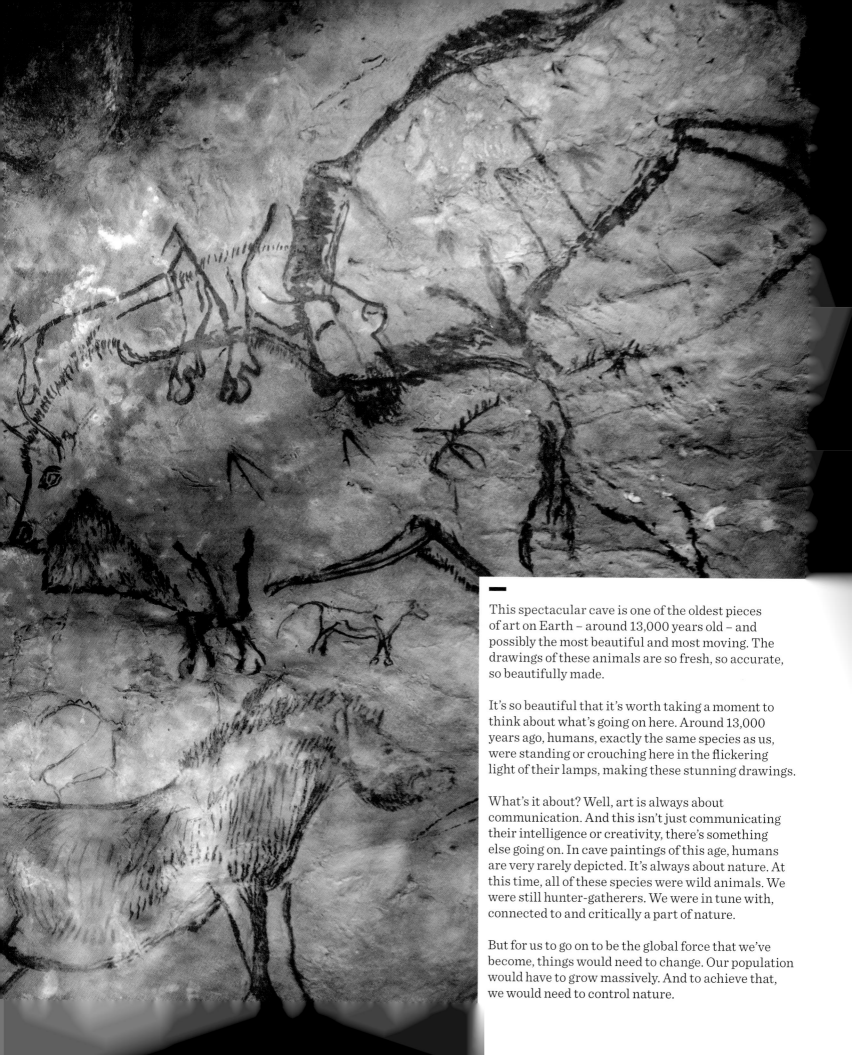

This spectacular cave is one of the oldest pieces of art on Earth – around 13,000 years old – and possibly the most beautiful and most moving. The drawings of these animals are so fresh, so accurate, so beautifully made.

It's so beautiful that it's worth taking a moment to think about what's going on here. Around 13,000 years ago, humans, exactly the same species as us, were standing or crouching here in the flickering light of their lamps, making these stunning drawings.

What's it about? Well, art is always about communication. And this isn't just communicating their intelligence or creativity, there's something else going on. In cave paintings of this age, humans are very rarely depicted. It's always about nature. At this time, all of these species were wild animals. We were still hunter-gatherers. We were in tune with, connected to and critically a part of nature.

But for us to go on to be the global force that we've become, things would need to change. Our population would have to grow massively. And to achieve that, we would need to control nature.

THEY
DIDN'T
LOOK UP...

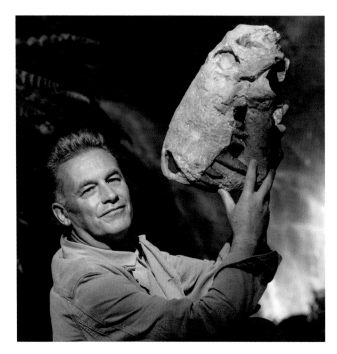

Opposite page: The dinosaurs were a diverse and charismatic group of animals that have long inspired and intrigued humanity.

Above: Coming face to face with the dinosaurs while shooting *Earth,* the series.

The final chapter of our story begins 66 million years ago in the final days of the Cretaceous Period, when the Earth was very much in full bloom. The climate is warm, the oceans are bursting with a multitude of life, and on land flowering plants have created a kaleidoscope of colour. And as has been the case for almost 150 million years, the dominant terrestrial creatures of the age are the diverse group of reptiles we call the dinosaurs. The skies are filled with giant pterosaurs like *Quetzalcoatlus*, one of the largest-known flying animals of all time, and on land 9-metre *Triceratops* can be found rummaging in the low-growing vegetation for the vast volumes of food they needed to sustain them. Lurking not far away are the great predators of the times, the theropods that included *Tyrannosaurus rex*, one of the largest and most fearsome land predators of all time.

Life seemed to be thriving in every corner of the Late Cretaceous world – the planet was ice-free and rich dense forests extended all the way to the poles, creating an Eden that looked as if it could go on forever. But as we know, forever is never part of Earth's constantly shifting story; this is an endlessly dynamic planet, within an endlessly dynamic universe, and whenever it looks as if stability has settled in, events always take over.

The event we are talking about is, of course, the asteroid that collided with Earth some 66 million years ago, and which, as we have now all heard many times, 'wiped out the dinosaurs'. This was about the most rapid and unexpected change that could happen on our planet. It was a day that would have begun like any other for the billions of life forms on Earth, but that ended with a shattering shift in direction for almost all life. The cause of this cataclysm was an asteroid estimated to be 10 kilometres in diameter, which struck the Earth at 20 km/second. The resulting impact crater is estimated to have been 180 kilometres in diameter and 20 kilometres deep – a vast hole literally punched into the surface of the planet. Now, extraordinary claims like this require extraordinary evidence, and an asteroid colliding with Earth and unleashing energy equivalent to something like a billion nuclear weapons is certainly an extraordinary claim! So, where's the proof that this really happened – where's the 'smoking gun'?

The answer to this begins in the late 1970s with the work of geologist Walter Alvarez and his father, Luis Walter Alvarez, a Nobel Prize-winning physicist who, like a twentieth-century scientific Forrest Gump, seems to appear in an endless stream of world-changing moments – from the invention of radar, to the Manhattan Project, to the assassination of JFK. A scientific detective through and through, there was no mystery too great for Alvarez to try to solve, including one of the greatest extinction events in Earth's history.

Alvarez Senior and Junior together were the first to put forward the theory that the extinction event at the end of the Cretaceous that wiped out the dinosaurs was caused by an asteroid impact. The key evidence they relied on to support this new theory was the discovery of a thin layer in the geological record containing unusually high levels of iridium. Iridium is an element that is rare on Earth but found in much higher concentrations in asteroids, so the discovery that the levels of iridium in this layer were 160 times above the background level and that the layer corresponded with the extinction of the dinosaurs, suggested a causal link. At first the theory was heavily questioned, but we now can say with great certainty that the iridium levels at the Cretaceous–Paleogene (K-Pg) boundary – a geological signature of thin bands of rock that mark the end of the Cretaceous and the beginning of the Paleogene Period – are direct evidence of the impact event. During this event, the asteroid vaporised on impact and spread iridium all over the globe, producing a layer that we have now discovered across the planet.

Below: Father-and-son research team Luis and Walter Alvarez have postulated that a giant asteroid hit the Earth in the Cretaceous Period, approximately 66 million years ago, and this caused an atmosphere like a nuclear winter, which possibly led to the extinction of dinosaurs.

Bottom: Illustration of a large asteroid colliding with Earth on the Yucatán Peninsula in Mexico. This impact is believed to have led to the death of the dinosaurs some 66 million years ago. The impact would have thrown trillions of tonnes of dust into the atmosphere, cooling the Earth's climate significantly, which may have been responsible for the mass extinction.

Right: A layer of iridium-rich rock, known as the K-T boundary, is thought to be the remnants of the impact debris of the asteroid that struck the Earth, causing the mass extinction event that killed the dinosaurs.

'In an asteroid impact,
the molten rock vaporises
to form a gas. That gas
expands to form a plume,
and inside the plume the
gas condenses, solidifies
and cools, and becomes
rounded droplets that
then rain down on the
surrounding environment.'
Jessica Watkins, NASA

Interest in the impact theory grew as more and more indirect evidence pointed to its accuracy, but like a murder without a body, there were no known remains of an impact crater anywhere on Earth that were the right age and size. This was a theory that still needed to find the grand location where it had all played out, so in 1981 scientists from across the world gathered in Snowbird, Utah, to discuss where the remains of that telltale crater might be. Unbeknown to the scientists at Snowbird, another much more poorly attended conference was going on that year, where the actual answer to that big question was being presented.

Geophysicists Glen Penfield and Antonio Camargo had been working for PEMEX, the Mexican state oil company, in the late 1970s on a series of surveys in the Gulf of Mexico, just north of the Yucatán Peninsula. Noticing a series of 'magnetic anomalies', they began to explore a particular feature in the area that appeared to them like a 'bullseye', a shallow structure 180 kilometres in diameter that had all the hallmarks of a massive impact crater. Corporate policy at PEMEX led to caution in releasing any data from Penfield and Camargo's research, but they did present their findings at the low-key Society of Exploration Geophysicists conference in 1981, with the publication of a corresponding paper. However, with little attention given to the work and all routes to further research blocked, Penfield and Camargo moved on and returned to their PEMEX assignments, while scientists around the world continued the search for the dinosaur-destroying crater.

It wasn't until the beginning of the 1990s that the significance of that discovery in the Gulf of Mexico finally came to light, when a journalist who had been aware of the initial discovery tipped off one of the leading crater-hunting teams of the time, led by Alan Hildebrand. Hildebrand made contact with Penfield and together they sourced core samples that had been drilled from the Yucatán site decades earlier by PEMEX teams. The results of the tests were astounding; the samples showed clear signs of the shockwave and heating associated with an impact event. Combined with new satellite data that revealed the outline of a clearly visible geological feature that matched the one Penfield had seen a decade earlier, the team, including Hildebrand, Penfield and Camargo, published a landmark paper. In it, they announced the site of the impact crater that changed the course of life on Earth. The location of the crater centred around the small town of Chicxulub Puerto.

Below: The aftermath of the Chicxulub asteroid impact killed all species of dinosaur, including this Thescelosaurus.

Below right: The location of the Chicxulub impact crater, in Mexico, where the asteroid struck.

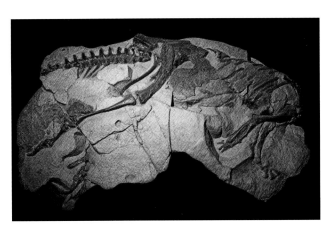

Right: The top picture is a
relief image of the Yucatán
Peninsula that shows a subtle
indication of the Chicxulub
impact crater. The bottom
picture is the same area
viewed in infrared by the
Landsat satellite.

CHICXULUB IMPACT
A schematic of the crater caused by an asteroid
striking the Yucatán Peninsula.

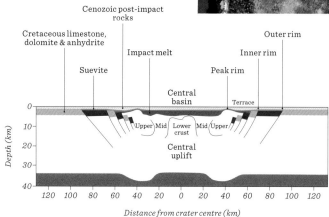

Opposite: There is a ring
of cenotes, or sinkholes,
surrounding the Chicxulub
impact crater, caused by the
asteroid's strike weakening
the rock. Many are popular
tourist destinations today.

Today, anyone lucky enough to visit the Yucatán Peninsula can visit the only tangible evidence still visible on land of that catastrophic day 66 million years ago. This evidence also happens to be in the beautiful form of a type of sinkhole known as a cenote. Cenotes are formed from the collapse of limestone bedrock and are most often water-filled, circular pits that are located in their thousands across the Yucatán. The word 'cenote' comes from one of the ancient Mayan languages and literally means 'hole in the ground filled with water', but despite their simplistic name, these are incredibly beautiful structures.

What makes these cenotes more than just another tourist-attracting geological feature only becomes apparent when you look away from the beauty of any single cenote and look at a map of them in this part of Mexico. There are an enormous number freckled across the landscape, but look closer and you can see that the distribution is not entirely random. Running across the north-western end of the peninsula is an unmistakable arc of cenotes, and when the semicircle is completed into a circle it has a diameter of around 180 kilometres, which corresponds perfectly with the scale of the inner rim of the crater that was made when the asteroid struck our planet. The cenotes weren't formed at the time of the impact; that impact fingerprint would take millions of years to appear, a geological feature created by the impact of the asteroid weakening the rock along the crater's edge. Over time water has slowly eaten away at the limestone in a particular pattern, forming cave systems that have collapsed in some places to give us this visible ring of cenotes. It is an extraordinary marker of the events of that day 66,043,000 (± 11,000 years) million years ago.

Remarkably, we are still piecing together more and more details about the events on that distant day. We think that the asteroid was of a carbonaceous chondrite composition, and at around 10 kilometres in diameter it was an object taller than the peak of Mount Everest. It came screeching into the Earth's atmosphere, approaching from the northeast what was then a very different coastline at an angle of around 50 degrees. On impact it collided with an ocean floor that was 100 metres deep at its shallowest point and 1,200 metres at its deepest. At the moment of impact, this vast rock was travelling at 20 km/second, with a kinetic energy on impact estimated to be equivalent to 100,000 gigatonnes of TNT. The blast from the asteroid impact annihilated everything within its reach, with winds reaching 1,000km/hr near the impact site, mile-high waves swamping the coastline and the triggering of earthquakes of an estimated magnitude of 9–11 Mw, as well as mega tsunamis over 100 metres tall that tore across the planet. The impact was so powerful it carried material from the impact site all the way to what is now Texas and Florida, reverberating through the Earth and leaving its seismic mark over 6,000 kilometres away. If that wasn't enough, the impact created a vast cloud of dust and ash that blasted what has been estimated to be 25 trillion tonnes of material from the impact site into the atmosphere. This in turn triggered a downpour not of rain but glowing-hot rock, cooking the atmosphere and driving it up to oven-like temperatures, and igniting a contagion of wildfires.

It was, without doubt, one of the worst days in the history of the Earth thus far, but even taking into account all of that overwhelming destruction, it would still not have been impactful enough to explain the levels of extinction that we know ultimately occurred, not just at the impact site but across the globe. For a species to become fully extinct, every member of it has to die and not be replaced, and not just in a small area or even over a continent, but across the entire planet. That's what extinction is really about, and almost always it is not a dramatic process but a long, slow, excruciating loss. So the question is, what made this moment so overwhelmingly destructive of so much life on Earth?

CASCADE OF EXTINCTION

To understand why the consequences of the collision were so widespread, let's take a look at one particular dinosaur that was doing very well for itself all across the planet before the asteroid hit. Ornithomimidae were a family of therapod dinosaurs that bore a remarkable similarity to the modern-day ostrich. Just like the ostrich they could run really fast, perhaps among the fastest of all dinosaurs on land – a useful trait to help avoid the attentions of predators in the Late Cretaceous. We are not entirely certain if they were pure herbivores or more omnivorous in their diet, with the occasional small prey thrown into the daily feast. What we do know is that the majority of their diet came from plants, because we have found many fossils of these creatures containing gastroliths (small stones) in their stomachs, which are a classic characteristic of herbivorous animals.

TEMPERATURE FLUCTUATION OVER THE LAST
500 MILLION YEARS

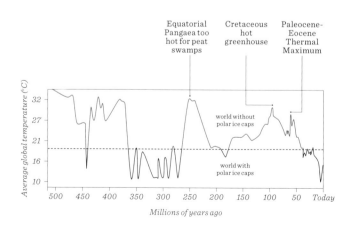

Equatorial Pangaea too hot for peat swamps

Cretaceous hot greenhouse

Paleocene-Eocene Thermal Maximum

world without polar ice caps

world with polar ice caps

Average global temperature (°C)

32

27

21

16

10

500 450 400 350 300 250 200 150 100 50 Today

Millions of years ago

'Anhydrite is very rich in an element called sulphur – and the sulphur is what's really important. When we drilled into the crater, the cores from the layers of rock where the asteroid actually hit revealed no anhydrite. We think this means all the anhydrite was vaporised by the force of the impact. This would have put about 325 billion tonnes of sulphur into the upper atmosphere, which is where this impact had its devastating effect.'
Chris Lowery, Jackson School of Geosciences, University of Texas at Austin

Opposite: *Gallimimus* (left) and *Dromiceiomimus* (middle) were dinosaurs in the Ornithomimidae family. They look very similar to the modern ostrich, but a direct relation has not been proven.

Around 66 million years ago, at the end of the Cretaceous, Ornithomimids were the most common small dinosaur in what is now North America, and could also be found roaming across large parts of Asia and we think Australasia, too. We know from the fossil record that on the day the asteroid hit this was a very successful dinosaur, living in vast numbers across much of the globe. So why did they become extinct? Well, for those unlucky enough to be living within the blast radius of the asteroid the answer is simple: they were vaporised instantaneously. For those living a little further away, say, within a thousand miles or so of the impact site, the earthquakes, mega tsunamis and fires would have made sure there were very few, if any, survivors. But what about the ornithomimids that were living in Asia on that day, grazing peacefully on the other side of the world, as far away as possible from the immediate destruction that was caused by that asteroid? They would have experienced nothing of the devastation, neither immediately on impact nor in the days afterwards.

To understand what killed these creatures, you have to understand the aftermath of the asteroid impact, a time when all across the planet darkness fell and temperatures plummeted as a new, omnipotent wave of death began to sweep across the Earth. This was a planet plunged into the grip of a bleak, cold, dark winter with seemingly no end in sight. But this was not any ordinary winter, this was an 'impact winter' caused by the vast amounts of material thrown into the upper atmosphere when the asteroid struck the planet; a toxic cocktail of soot and dust from the billions of tonnes of rock that were vaporised in the impact, thrown into the atmosphere and blocking out the Sun. There was so much material sent up into the atmosphere that it could have covered almost the entire surface of the Earth. But if that wasn't bad enough, we are now beginning to understand that the deadly circumstance that the impact created was compounded even further by a particular quirk of fate.

In the 1950s petroleum engineers were surveying for oil in the Yucatán when they found large amounts of a particular type of rock beneath the surface of this part of the peninsula. Anhydrite, or anhydrous calcium sulphate, is a mineral with the chemical formula $CaSO_4$ that has been found across the planet, but it is more commonly found in its hydrated form ($CaSO_4 \cdot 2H_2O$), otherwise known as gypsum. Depth is critical for anhydrite to persist; too near the surface and it will inevitably come into contact with groundwater and so be altered into the hydrated form gypsum. It means that anhydrite is relatively rare in the Earth's crust, but when scientists estimated its presence in the marine sediments of the Yucatán from 66 million years ago, it seems that it may have made up to a third of the seafloor.

It's this abundance in this particular spot that makes the fact that the asteroid impact happened here all the more of a catastrophic coincidence, because when the asteroid struck, it vaporised this rock, releasing no less than 300 billion tonnes of sulphur into the atmosphere. The effect of this sulphur was perhaps the ultimate tipping point that turned a localised extinction event into a global one. The soot and dust that the asteroid threw up into the atmosphere was damaging to life all over the globe, but we think this would have rapidly rained back down onto the surface in a matter of months or, at worst, years.

All of that acid rain falling on the Earth was far from good, but it wasn't as destructive as what was going on in the upper atmosphere, because all the sulphur released from the anhydrite behaved in a very different way, hanging around for decades in the upper atmosphere, blanketing out the Sun and prolonging the effect of that agonising impact winter. It seems that the real killer at that time was not the bitter cold but the near complete darkness caused by the sulphur, which reduced the amount of sunlight reaching the surface of the Earth by more than 50 per cent.

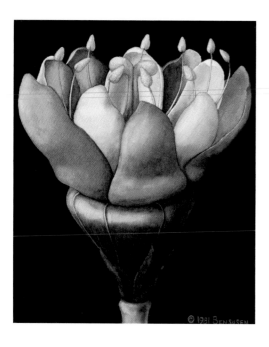

Below: Illustration of a fossilised angiosperm from around 80 million years ago that is thought to be a cousin of the rose.

With so little sunlight, plants struggled to photosynthesise, creating massive levels of disruption and ultimately extinction to terrestrial flora. Studying the fossil record of pollen and leaf remnants of the time suggests a mass dying that in North America alone resulted in over 50 per cent of plant species becoming extinct – and even in areas of the globe where the conditions didn't drive species to extinction, they did cause a mass downturn in the abundance of plant life. This of course meant that any species that depended directly on an abundance of plant life began to suffer. For those herbivorous dinosaurs that had survived the impact, it was now a planet that was starving them to death, and once the herbivores begin to dwindle, the carnivores soon began to starve as well. With diminishing sources of food, they too faced an ecosystem that could no longer sustain them. In this environment, flexibility and opportunity were precious characteristics favouring the omnivores, insectivores and scavengers far more than the pure herbivores or carnivores. Size was also a disadvantage, because those giant dinosaurs were simply too big to be sustained by the new energetics of this shadow Earth, so as a general rule any four-legged creature heavier than 25 kilograms didn't make it through.

Across the planet, on every continent and in every ecosystem, life suffered in the midst of this deteriorating environment, with estimates based on marine fossil records suggesting 75 per cent or more of all species of animals and plants on Earth simply disappeared. When it came to the dinosaurs, the impact was even greater, with only a handful of smaller species hanging on to evolve into the 10,000 or so species of birds that we have today. This was one of the worst extinction events our planet has ever seen, and yet, even amid the incredible devastation brought by the impact, some species managed to survive.

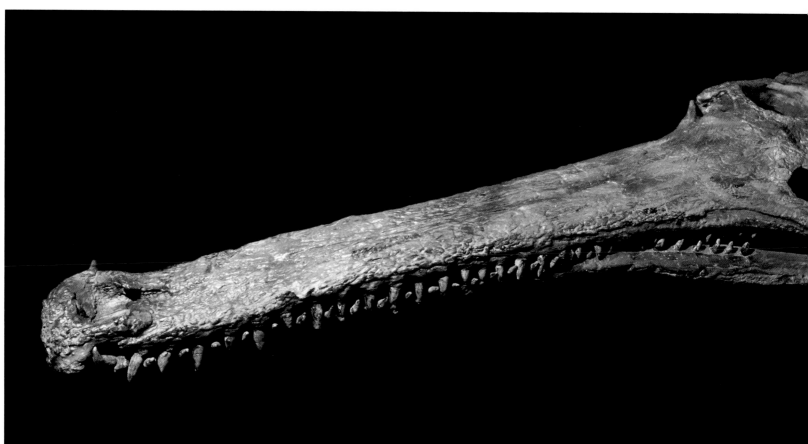

Surrounded by an abundance of decaying vegetation, fungi thrived, as did insects and other creatures like worms that relished the growing layer of detritus. In the water, the scarcity of food favoured crocodilians, with their small, slow-growing young and ability to last months without a reliable source of food. And crucially for us, on land it was small, opportunistic omnivorous mammals that seemed to survive in these conditions. They certainly didn't make it through the extinction event without heavy losses, though, and in the aftermath they didn't thrive straight away either. But small in size, adaptable in sustenance and sheltered away from the maelstrom with many species in burrows, they emerged into a world where the opportunity for mammals had completely changed. When the skies finally cleared and the Sun shone down on the Earth, it was the small furry mammals that must have looked up with a new sense of hope.

Left: Fossil jaw of the prehistoric crocodile species *Dyrosaurus phosphaticus*, which dates back to about 65 million years ago.

Below left: *Terminonaris robusta*, a prehistoric crocodile that reached lengths of 6m.

Below: As the dinosaurs and the majority of all plant and animal species on Earth died out, small mammals thrived in the vacuum.

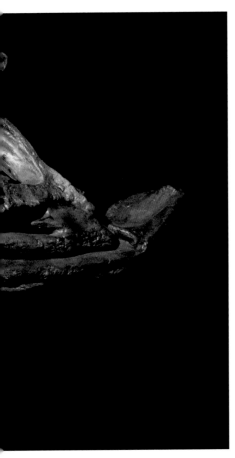

Sixty-six million years ago, the mammals didn't all become extinct – some species survived and were uniquely suited to the chaotic new world, which was essentially covered in the detritus of the old. Long, twitchy whiskers. Those lovely bright eyes. Our own ancient ancestor likely shared significant similarities with creatures like the modern rat. So what did they have that the dinosaurs didn't?

The dinosaurs were quite specialised. The herbivores required very particular species of plant to eat, while carnivores relied on other animals for their food supply. The early mammals like the rat, though, were generalists and, equally importantly, they were omnivores. They could eat plant material and flesh. In fact, they could eat rotting plant material and rotting flesh. All of those millions of years ago animals like this represented hope, and the world that we know today was resting on the shoulders of creatures just like this rat.

NEW
DAWN

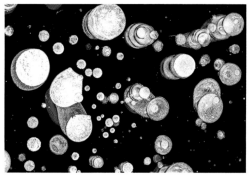

The reign of the dinosaurs, which had dominated Earth for 165 million years, had come to an end and a new chapter for life was now beginning. This was the Paleogene Period, an era in the Earth's history that would last 43 million years and lead us from a landscape of post-apocalyptic destruction towards a new age that would be defined by the rise of the mammals. But, as always, this would be no straightforward journey for the Earth and all the life that lived upon it, because just as the planet's ecosystems began to recover from the impact, Earth's climate began to radically change, with the most rapid and significant climate-warming event of the past 65 million years. Known as the Paleocene–Eocene Thermal Maximum, or PETM for short, this was a moment in the Earth's history around 56 million years ago when global average temperatures soared by as much as 8 degrees Celsius in a period of global warming that lasted for about 200,000 years. It's an event that is of substantial interest to scientists looking at the current global warming spike we have created and are living through right now, because within that distant event are clues to how the Earth reacts in moments like this and how tipping points that trigger a cascade of events can be reached.

The cause of the PETM, unlike today's climate crisis, was nothing to do with a rogue life form releasing vast amounts of greenhouse gases into the atmosphere. Back then the planet once again seemed to turn on itself. The trigger for this event was hiding deep beneath the North Atlantic Ocean in an area known as the North Atlantic Igneous Province, centred around what is now Iceland. Deep beneath the ocean hundreds of volcanic vents were erupting, transforming the submarine landscape and in doing so releasing vast quantities of methane, which, like carbon dioxide, is a greenhouse gas, only it's twenty-five times more potent. On the seafloor all of this activity created a vast basaltic lava plain ranging over 1.3 million kilometres and known as the Thulean Plateau. Stretching across what is now the North Atlantic, at one point in Earth's history this structure formed a North American land bridge that extended from Greenland to the British Isles, connecting North America with Eurasia. Now, that would have been quite a walk!

However, for all the action that was happening under the water, it was in the atmosphere that the real transformation of our planet was taking place. As the methane rose to the surface and filled the atmosphere, it kick-started an already warming Earth into overdrive. And it didn't stop there. Just as we are concerned about our own modern-day climate crisis, the increase in temperatures began a ripple effect of other events, crossing a tipping point in the planet's ability to regulate its temperature. The release of vast deposits of methane locked within the seabed caused them to rise upwards like a 'giant planetary burp' and push temperatures even higher, transforming the environment for any creature that had survived – including our distant ancestors, those early mammals.

Top left: Hot lava from Kilauea volcano in Hawaii, USA, erupting under water as pillow lava, forming new oceanic crust.

Middle left: Methane rising up from the seabed.

Left: The Giant's Causeway in Northern Ireland is the result of an ancient volcanic fissure eruption.

Opposite: The American alligator and all other crocodilian species are living relatives of the dinosaurs.

'The warming event had an impact on virtually every environment on Earth. With the combination of warming and ocean acidification, there was one of the largest deep-sea mass extinctions in recent Earth history.'
James Zachos, University of California Santa Cruz

With all that heating came dramatic changes to the weather. As average global temperatures reached as high as 24 degrees Celsius, the rates of evaporation of water from the oceans increased rapidly, driving it up into the atmosphere, making it not just a warmer but a much wetter planet, too. This water vapour was not found just in the places on Earth that we would expect; studies have shown that much of the additional moisture was transported towards the poles, where the once icy caps of our planet were transformed into subtropical havens. Around 56 million years ago, the global warming event was so extreme that alligators could be found lounging beneath palm trees in the Arctic. As we have seen throughout the history of the Earth, massive change – in this case rapid climate change – results in a period of disruption and upheaval that has both winners and losers, and in the case of the PETM, one of the winners appears to have been our distant ancestors. The tropical conditions seem to have been a good thing for those small creatures scurrying around, because for the first time a very special environment was allowed to expand and flourish. This was the birth of the modern tropical forests, with all their plant and animal diversity, and it was this rich habitat that would prove to be crucial for the evolution of our mammalian ancestors.

CLEVER PLANTS

Fifty-six million years ago our planet was a hot and humid tropical world with vibrant, steaming forests covering entire swathes of North America, Europe and Asia. Within these forests were the blooms of endless flowering plants, a feast of colour that would provide mammals with an extraordinarily precious gift. To understand this gift, we first need to consider why flowering plants – or angiosperms, as they are known – are so fundamentally different to other seed-producing plants. The term 'angiosperm' is derived from the Greek words *angeion*, meaning container, and *sperma*, meaning seed – a name that describes that other characteristic feature of flowering plants: fruit. Fruit plays a crucial role in the life of flowering plants, emerging in myriad varieties from endless forms of flower. The fruit of an angiosperm develops after the pollination of the ovule deep within a flower, with the ovule becoming the seed and the ovary within which it sits, while other parts of the flower develop into the fruit. A structure that has evolved to protect the fertilised seed of a plant and often helps in the germination of the seed, the fruit can also play a crucial role in the dispersal of seeds away from the parent plant by attracting animals to come and eat it.

This is where a mutually beneficial relationship with mammals comes into play. For the plant, it's a perfect way to spread your progeny, by having an animal eat a seed and then carry it off from the parent plant and deposit it in its faeces, ready to

Below: In the Earth's new climate, angiosperms evolved alongside their pollinators and flowering plants like this bromeliad spread across the world, transforming the planet with a riot of colour.

Below: *Rafflesia* flowers look and smell like rotting flesh to attract their pollinators, carrion flies.

Bottom: Bumblebee orchids trick insects into pollinating them, thinking they have found a mate.

Right: Unlike most flowering plants, the clock vine is pollinated by birds and mammals rather than insects, in this case, the Pallas's long-tongued bat.

grow some distance away. On the other side of this cooperative relationship, for the large number of mammals that evolved to eat fruit, it offers both a great reward but also a challenge. The reward is obvious – a perfectly formed, fruit-shaped capsule packed full of calories – but basing your diet on fruit is not necessarily easy, because once you commit to being a frugivore you've not only got to find the fruit, you've then got to be able to reach it, and in a forest environment that's actually quite difficult. To find fruit you've got to be able to see it among the foliage and then identify whether it's ripe or not. Then once you've done that, you most likely have to be able to climb 20 or 30 metres to pick any of it.

Left and below: New World monkeys became a distinct group around 40 million years ago, evolving from the Old World monkeys that colonised South America.

Below: Geoffroy's spider monkey is a New World monkey from the family Atelidae, which have evolved fully prehensile tails.

Left: *Purgatorius* is a genus of seven extinct species believed to be the earliest examples of proto-primates, dating to as far back as 66 million years ago.

EVOLUTION OF PRIMATE GROUPS

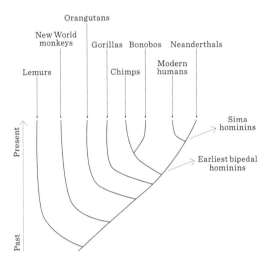

'With this global temperature spike, forests started to expand further and further, even up into the northern hemisphere. And with that, primates were also then able to expand into Europe, North America, Asia and Africa.'
Ammie Kalan, University of Victoria

That's why the nutritional benefits of fruit, combined with the challenge in accessing it, became an evolutionary driver that would lead to the emergence of new types of mammal far more dynamic than those ancient rat-like creatures. Today we see the very existence of these forest-dwelling creatures being threatened all over the world; creatures like the Geoffroy's spider monkey, one of the largest monkeys native to Mexico and a truly amazing animal. These strange and beautiful monkeys live in groups of up to 40 individuals, existing on a diet of ripe fruit and leaves that requires them to traverse large tracts of forest to survive. You only have to look at them to see how specialised they are for this environment, with a set of anatomical features that are tuned to maximise the gastronomic opportunities of the forest. Their eyes are forward-facing like ours, which is good for stereoscopic vision and helps spot the ripest, most delicious fruit when they're on the move. They've also got relatively big brains so they can identify different types of fruit and remember where they previously found it – in fact, one study suggested they were the third most intelligent non-human primate after orangutangs and chimpanzees. But perhaps their most pronounced adaptation to the forest is the long, strong limbs and hook-like hands that allow these monkeys to swing so effortlessly from branch to branch. And if the arms get tired, well, they can always rely on their fifth limb – the prehensile tail that is so strong it not only helps them move but is also able to hold their full body weight while they hang from a branch and feed.

To see these creatures in their natural habitat leaves you with no doubt that they are perfectly adapted for the high life at the top of the forest, and many of these adaptations can be traced back to their distant ancestors, those first true primates that arose around 65 million years ago. Exactly when and where the first primates emerged is not entirely clear, because as soon as they evolved, they started spreading rapidly around the world: across North America and Asia and on into Europe and Africa. One of the earliest proto-primates to begin this new lineage may have been *Purgatorius*, an animal that looked more squirrel than monkey but might have been a precursor to the multitude of early primates that we know emerged in North America around 55 million years ago. It would take another 15 million years before the first primates that we would recognise as monkeys and apes would appear. These simian primates may have first emerged in Asia but they soon spread to Africa, South America and beyond.

This was a boom time for primates all over the globe. They were rapidly spreading and diversifying, and it's perhaps the closest our world has ever got to being a true Planet of the Apes. But it wouldn't last. As the Earth's climate slowly cooled, the habitats that had driven the explosion of primate life began to disappear, the spread of tropical forests retreated, and with that our distant ancestors were about to be pushed onto the back foot.

As the tropical forests disappeared from northern climes, the consequences were far-reaching. Primates disappeared entirely from North America and all but disappeared from Europe as well, in what at first glance seemed to be a serious setback for our ancient ancestors. But as we have seen again and again in Earth's long history, a setback for one form of life is often an opportunity for another, and if the story we're *really* telling is the story of us – *Homo sapiens*, humans – then this shift in environment was one of the most critical turning points in our family history.

The path of evolution that would lead from those ancient ancestors to a creature like us – an upright, bipedal, big-brained ape with opposing thumbs – was never going to be straightforward. For evolution to generate something like the human species, it would require a very precise and complex series of events, and it would need to occur in a very specific environment where the evolutionary drive would be to live not up in the trees but down on the ground.

The Calakmul Biosphere Reserve in Mexico, with its 7,000 square kilometres of forest, is one of the largest forest reserves in Mexico and Latin America. Aside from Mayan ruins and a few other pockets of development, there's very little human activity here, which means it's packed full of wildlife – with more than 400 species of birds: toucans, trogons, parrots and tanagers. It's not just birds, though. Hiding in the forest here are all sorts of other natural wonders: ocelot, puma, monkeys, enormous numbers of bats, not to mention a host of reptiles and amphibians.

So the question is, why is this place such a biodiversity hotspot? It's partly due to the vast diversity of plants here – we think that in this area alone there are more than 1,500 species – but it's not just that. It's also down to something that we very much take for granted.

Most of the plants here will produce flowers – some big, some small – and because of this characteristic we call them 'flowering plants' or angiosperms, and when it comes to influencing biodiversity, it's these angiosperms that are really important.

ON TWO LEGS

Around 34 million years ago, primates had mostly disappeared from the cold, dry northerly continents, but in East Africa, where an abundance of tropical forests still remained, they had survived. And this is where the forces of the planet conspired to create a new environment that would dramatically change the direction of life on Earth, as the continental plate of Africa began to pull itself apart.

Today we see the result of this tectonic fracturing in the rift that runs some 7,000 kilometres from Lebanon in the north to Mozambique in the south. Around 35 million years ago, when this great East African Rift System first began to open up, its impact was widespread, creating a new environment where the once all-conquering forests had become a patchwork, with vast savannahs in between. In evolutionary terms this radically new environment presented our distant primate relatives with a tricky decision. To understand this dilemma you have to imagine for a moment that you are a highly arboreal, tree-swinging, fruit-eating primate and you've had your breakfast in forest number one but your lunch is over there in forest number two and there is a big bit of grassland in between. So, how do you get between the two as quickly as possible without being taken out by a predator? When your environment is entirely forest, the answer to that question is simple: you swing through the trees to get from the first forest to the second. But with grassland the challenge is different, and this was the dilemma facing those early primates. For some, the decision was to stick it out in the shrinking forest, and it was these primates that evolved into chimpanzees and gorillas. But another group – our ancient ancestors, the hominids – started to do something that was quite literally a revolutionary step in evolutionary terms: they started walking on two feet. Exactly when and why our ancestors first started doing this is still one of the many great mysteries that remain hidden in our distant past, but there is little doubt that it eventually gave those upright ancestors a very distinct advantage.

We now think that bipedalism is in fact the most fundamental characteristic that separated the first hominids from the rest of the apes that continued to walk on four legs. We used to think that it was brain size that made our hominid lineage unique, but at the turn of the twentieth century, as we looked at the skulls of the

'This was really an evolutionary fork in the road for primates. Food resources became more dispersed which meant that primates had to travel further in order to find enough food to survive. And what we see is that this likely led to evolving more efficient ways of moving through the landscape.'
Ammie Kalan, University of Victoria

Below: A trail of hominid footprints fossilised in volcanic ash. The trail was found by Mary Leakey's expedition at Laetoli, Tanzania, in 1976. It dates from 3.6 million years ago and shows that hominids (probably *Australopithecus afarensis*) had acquired an upright, bipedal, free-striding gait by this time.

only early hominids that we had as yet discovered, such as *Homo erectus* and Neanderthals, it seemed that we were part of a group of big-brained apes that had thought our way out of Africa. While there is, of course, some truth in this, the discovery of a 2.3-million-year-old skull in a limestone quarry in 1924 fundamentally challenged that perspective.

Uncovered by a group of quarrymen working in the town of Taung, in the north-west province of South Africa, the Taung skull, as it came to be known, would transform our understanding of early hominids and ultimately of our own evolution. This was a primate skull, a long-extinct species of monkey with a small brain and human-like teeth, but what made this discovery immediately controversial was the particular shape of one part of its anatomy. The foramen magnum is the passage along which the spinal cord passes through the skull and joins the brain, and in humans this is positioned in a very different way to other apes, lying further forward and directly under the skull on a line that runs straight into the spine. It's one of the crucial anatomical features needed to hold our head and spine in a position that lets us walk upright, and it was this human-like positioning of the foramen magnum in the Taung skull that was so surprising, because it seemed to suggest that this small-brained monkey walked upright at least some of the time. Although it took decades for this theory to be fully accepted, the discovery of this and other similar skulls led to its classification as an extinct species of hominid known as *Australopithecus africanus*, which lived between 3.3 and 2.1 million years ago in the south of Africa.

Since that discovery we have been able to trace the origin of bipedalism even further back in time through a series of discoveries, including the 4.4-million-year-old *Ardipithecus ramidus*, a species of australopithecine from the Afar Region of Ethiopia that shows clear evidence of its ability to both walk on two legs and live in the trees. Even older than that is *Orrorin tugenensis*, a species of hominid whose relation to modern humans is not entirely certain but whose thigh bone and teeth suggest that this was a creature walking on two legs at least some of the time, 6 million years ago. Oldest and still most controversial of all are the partial remains of a skull of a homininae discovered in northern Chad in 2002. Nicknamed

Above: Ethiopia's Great Rift Valley, where the first hominids walked, is called 'the cradle of humanity'.

Opposite left: The Taung Child skull came from a bipedal hominid species which walked 2.6 million years ago.

Opposite right: Human and orangutan skulls, compared in 1848 in *The Natural History of Man* by James Cowles Prichard.

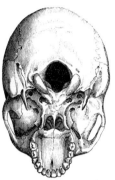

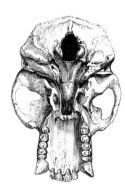

Base of Human Skull.   Base of Skull of Orang.

Toumai, at 7 million years old it's the oldest fossil from a member of a human family discovered to date. The remains of an extinct species known as *Sahelanthropus tchadensis*, this creature may have been alive at a time when humans and chimpanzees were moving down separate evolutionary pathways, making it a distant ancestor of both us and our chimp friends. Once again, the foramen magnum in this skull is positioned in a way that suggests it spent at least some time on two legs, but subsequent discoveries mean that the exact nature of its bipedalism are still being debated.

Wherever and whenever our early ancestors first walked, what we do know is that they didn't just stride out of the forest never to return. The earliest hominids were designed for both arboreal and bipedal living – still spending a lot of time in the forest and probably sleeping in the trees. However, walking began to give them an advantage that would be unrelenting, providing a freedom to roam that would ultimately lead from being a defensive to an offensive trait. As the bipedal anatomy became more and more honed, it allowed those ancient ancestors of ours to turn from prey to predator, developing hunting skills that would allow their diet to expand to include a reliable supply of meat. With that change in nutrition came the energetics to support bigger and bigger brains, and so it was in the Rift Valley in East Africa, the 'cradle of humanity', that our early ancestors stepped away from their primate brethren and began the long, slow, but relentless journey to the global dominance we see today.

With hindsight, we simply have to marvel at the remarkable set of factors that came together to produce that highly unlikely chain of events. The seemingly impossible had happened; after the best part of 66 million years, during which time continents had torn themselves apart, mountain ranges had formed, the Earth's climate fluxed and flipped between hot and cold, a new force was about to be unleashed upon the planet. Humans had arrived and there was no turning back.

Left and opposite:
The family Hominidae, known as the great apes, includes gorillas (opposite), chimpanzees (above left) and orangutans (left), as well as bonobos and, of course, us.

Above: Lemurs resemble other primates, but evolved independently from monkeys and apes. They diversified hugely on Madagascar and now comprise more than 100 species. The Verreaux's sifaka, pictured above, has a galloping bipedal gait unique to sifakas.

THE HUMAN FAMILY

'To be able to survive in a particular environment, one person's knowledge isn't enough. Humans live in communities, and we share our knowledge. It's not one mind, but many minds working together. It's this sum of all of the knowledge that we have, that we then pass on from generation to generation, that archaeologists call 'cumulative culture'. This is key. This is what makes humans unique. And it's what has allowed us to move into all different kinds of environments and essentially live in every corner of this planet.'
April Nowell, University of Victoria

Below: Human ancestors and relatives, from left to right: *Adapis* (lemur-like, 50 MYA); *Proconsul* (primate, 23–15 MYA); *Australopithecus africanus* (3–1.8 MYA); *Homo habilis* (2.1–1.6 MYA); *Homo erectus* (1.8–0.3 MYA); and 2 *Homo sapiens* (from 92,000 and 22,000 years ago).

Opposite: The 23 paired chromosomes of a human male. (A female has XX rather than XY in the bottom right corner.)

As we have seen again and again, the story of Earth (or humans, for that matter) is never a simple one. With the emergence of bipedal primates around 7 million years or so ago, we made the final break from our primate relatives. We'd already split from gibbons, orangutans and gorillas long before, and now this last division between the human and the chimpanzee–bonobo lineages was marked by another fundamental change to our biology. This was the moment our genome was reshaped with the joining of two chromosomes (to form our chromosome 2), leaving us with one less chromosome than all the other apes: a bipedal ape with 23 chromosome pairs and a big future ahead.

From this moment on, the family tree of humanity would blossom in myriad directions that we are only just beginning to understand. The evolutionary journey to *Homo sapiens* was far from any kind of straight line; it involved a multitude of branches and interbreeding between closely related species that means we carry the past of many of these ancestors in our own DNA. So, let's have a quick (and rather simplified) canter through some of those 4 million years of human evolution and meet some of the family. We'll start around 4.2 million years ago with *Australopithecus*, an early hominin that through multiple species (including the *Australopithecus africanus* we have already met) spread far and wide across the continent of Africa. It's *Australopithecus* that we know gave rise to the genus *Homo*, with the earliest fossils we've found dating back 2.8 million years. Whether these remains are of our direct evolutionary ancestors or just close relatives of modern humans, we still don't know. But what we do know is that within the space of half a million years species like *Homo habilis* and *Homo rudolfensis* were developing complex cognitive skills that were transforming their way of life in both tool-making and hunting, leading to a diet that was becoming increasingly rich in meat. Next came *Homo erectus*, perhaps the most successful of all the hominin species to come before us, and the first that we would undoubtedly recognise as human.

Long-limbed, with a flat face and sparse body hair, *Homo erectus* was a skilled hunter able to bring down large animals in the hunt and may have been the first of our ancestors to utilise fire. These were complex social creatures, technologically advanced, with some form of early language, a culture that developed and lasted for almost 2 million years (to put that into context, we've been around for just about 300,000 years) until the last known population of *Homo erectus* we know of disappeared from what is today Java, around 110,000 years ago. It's a sobering thought that one of our most successful ancestors can spread so far and wide, flourish for so long and yet still ultimately end up extinct. The cause of that extinction is worth noting to be at least in some part due to a changing climate, placing this species out of sync with its environment and, ultimately, fatally

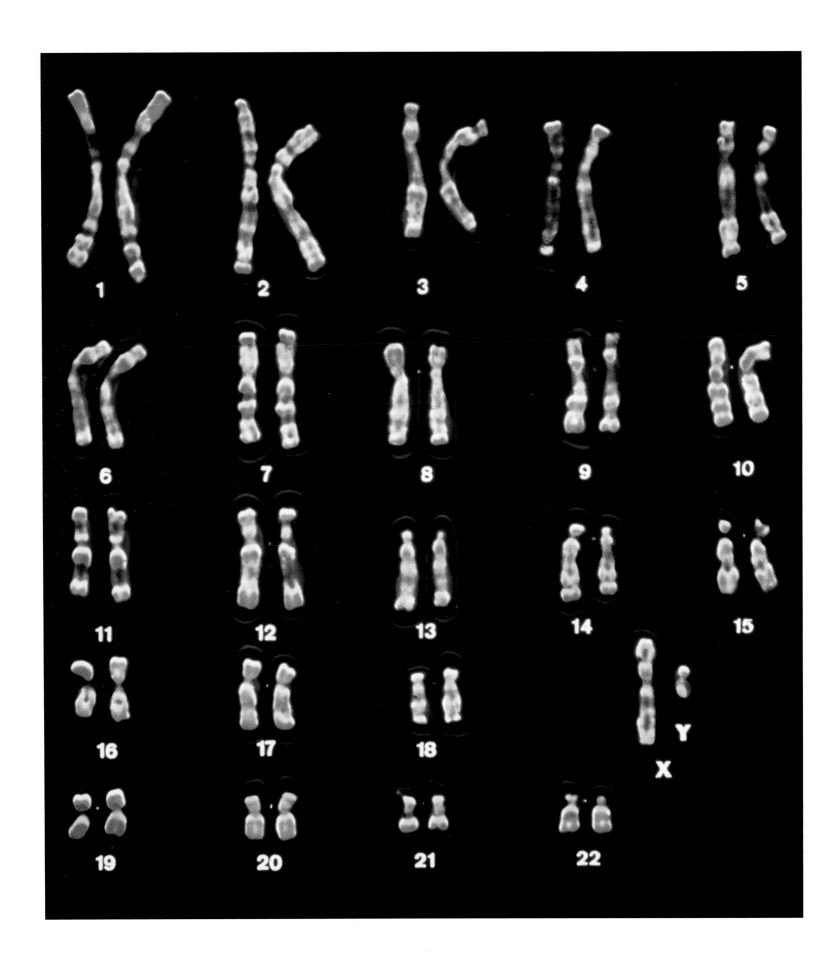

exposed. Somewhere along the long lineage of *Homo erectus* another species of hominin emerged that would ultimately lead to us. A number of candidate species have been discovered that may be the missing link between *Homo erectus* and *Homo sapiens*, but the exact hereditary line is still not fully known. What is certain is that around 300,000 years ago our species emerged in East Africa and at first lived a tentative existence with no sign of the global domination to come. But around 130,000 years ago something changed in our ancestors' behaviour and the first of at least two major waves of migration began the human journey out of Africa. Within the space of about 75,000 years *Homo sapiens* had colonised most corners of the world, including Eurasia and Australia, ultimately reaching the Americas around 15,000 years ago.

Left: This 1.5-million-year-old footprint found in Kenya belonged to *Homo erectus*, the most widespread of the hominids (with the exception of modern humans). *Homo erectus* lived between 1.8 and 0.3 million years ago.

Above: Cave art in the Andalusian cave of Ardales, in Spain, was painted between 43,000 and 65,000 years ago, 20,000 years before modern humans arrived in Europe – it was painted by Neanderthals. *Homo neanderthalensis* lived until about 40,000 years ago, and interbred with *Homo sapiens*, so we carry their genes to this day.

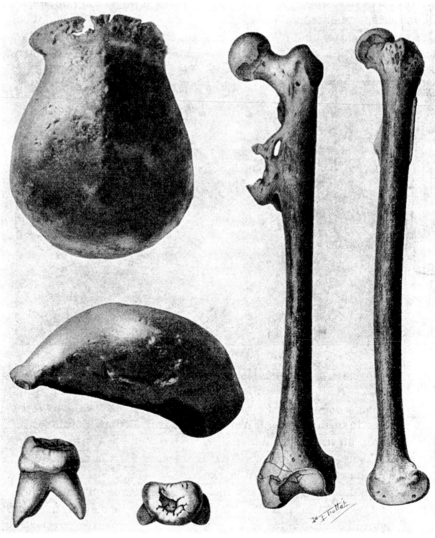

Opposite: Historical sketch of bones from Java Man, a *Homo erectus* fossil. These remains were found in 1898 in Java, Indonesia. *Homo erectus* were the most widespread and longest-surviving of all the fossil hominids. Their geographical spread included north and east Africa, Europe, Indonesia and China. They lived between 1 and 2 million years ago.

Right: Sunrise in Java, where the last *Homo erectus* population lived.

RESHAPING THE WORLD

Over the 3 million years that it had taken for our human species to evolve and begin that journey outwards to all corners of the globe, the Earth had also been evolving and transforming itself. Coinciding with the rise of our species, the Earth's climate was changing radically. Beginning around 120,000 years ago, not long after those first humans began to leave the warmth of Africa, the planet entered a period of dramatic cooling. Known as the Last Glacial Period (LGP), although we more commonly refer to it as the Last Ice Age, it was a period of cooling that would last over 100,000 years, reaching its peak around 22,000 years ago.

The LGP is part of an ongoing series of climate shifts known as the Quaternary Glaciation, which have seen the Earth bounce between ice ages over a period of two and a half million years, and 22,000 years ago the Earth was in the grip of the last of these glacial periods. For those humans that had made it across the globe, this was an Earth with an increasingly challenging and hostile environment. The northern hemisphere was most heavily affected, with Canada almost completely covered in ice sheets that extended south across the northern states of the USA all the way down to New York. Much of northern Europe, including the UK, Germany and Poland, was also covered in ice sheets that stretched eastward across Russia and towards China. At this point no less than one quarter of the Earth's land surface was under the ice, and even on the land that wasn't under ice the environment had become much more hostile. It was drier, colder and experienced a dramatic loss of vegetation due to the changing climate, and all of this was adding up to create an environment that was making it a lot more difficult for our species to spread around the planet.

But remarkably, even as we endured the cold, harsh extremes of this Ice Age, or perhaps because of it, we were beginning to adapt in extraordinary ways, developing a range of unique abilities. And these adaptations gave us an advantage not just over other primates but also over the other human-like species, such as the Neanderthals, that shared the Earth through that last Ice Age. By the time the ice

Below: Cave paintings dating from the late Neolithic, Epipaleolithic and early Bronze Ages depict humans hunting and herding animals.

Above: The goat was one of the first animals domesticated by early humans, perhaps only second to the wolf.

Right: Early humans were living and hunting alongside the ancestors of the first dogs 30,000 years ago.

1903

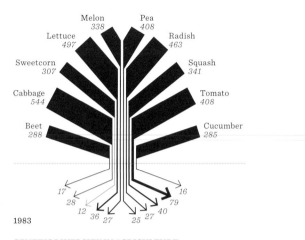

Melon
338

Pea
408

Lettuce
497

Radish
463

Sweetcorn
307

Squash
341

Cabbage
544

Tomato
408

Beet
288

Cucumber
285

17 16
28 79
12 40
36 27 25 27

1983

GENETIC DIVERSITY IN AGRICULTURE
This diagram shows the dramatic reduction in crop diversity over the last century. For example, while we used to regularly eat more than 300 varieties of corn, now we consume only 12. This reduction is due to the industrialisation of farming, and while it helps to produce food in mass quantities, it makes our society more vulnerable to crop collapse, and it has a negative impact on both human nutrition and ecosystems.

'With farming we transitioned from using the environment to owning the environment through domesticating animals, but also having a permanent landscape that we control. You don't just depend on what nature offers, you alter nature itself and adjust it to your needs.'
Professor Zeray Alemseged, University of Chicago

Opposite top: An example of forest gardening by the Kubu indigenous people in Sumatra, Indonesia. Immediately after clearing, rice is planted, followed by bananas and cassava in later years. This small-scale, varied type of farming is beneficial for pollinators and for the soil.

Opposite bottom: In contrast to a forest garden, an industrialised monoculture soya bean farm supports very little life due to the use of pesticides and lack of ecological niches for other species.

began to retreat around 13,000 years ago, it seems we had become the last human species still standing, and not just as survivors, because it was at this moment that human culture really began to flourish.

Around 12,000 years ago, as the climate warmed and the ice retreated, our distant ancestors, the hardy survivors of that ice age, unveiled a revolutionary idea, one that would change our relationship with nature forever and usher in a new age of humanity.

Like many of the early great leaps of humanity – be it the invention of the wheel, the use of fire to transform food through cooking, or the birth of agriculture – the exact origins and innovators remain lost in time. But what we do know about the birth of agriculture is that those first steps to domesticating a variety of plants and animals seemed to happen independently in multiple locations around the world. It began, we think, approximately 12,000 years ago in a strip of land across the Middle East in what has become known as the fertile crescent. Evidence of this first agricultural revolution, the Neolithic Revolution, has been found by archaeologists across this strip of land that stretches from eastern Turkey to southwestern Iran.

Exactly why the domestication of plants and animals began here we do not know, and the likelihood is we will never get the full picture behind the emergence of farming here. Multiple overlapping theories exist, but it's likely that climate change at the end of the last ice age brought with it a seasonal cycle that suited wild cereals like wheat and barley. This encouraged communities to fall into a more stationary annual cycle, with these plants as a source of nourishment, rather than continuing with the hunter-gatherer lifestyle that had existed before. Elsewhere, such as in Southeast Asia or South America, the circumstance driving humans towards farming may have been different but the outcome was the same; traditional hunter-gatherer lifestyles were replaced by permanent settlements coupled with a more reliable supply of food. And once humans had stopped moving, the hunger for agricultural innovation exploded. We've found evidence of fig trees being planted in the Jordan Valley over 11,000 years ago, squash being cultivated in Mexico 10,000 years ago, and a lot of genetic detective work has revealed that most lifestock, including cattle, goats, pigs and sheep, has its origins of domestication over 13,000 years ago in that fertile crescent.

It was this explosion of innovation that took us out of the Stone Age and propelled us into the modern era, because it was on the foundations of agriculture that cities, and ultimately civilisations, could be built. The results of this innovation have, as we well know, been transformative not just for our species but for the Earth as a whole, taking it from a place of endless biodiversity to a world filled with monocultures (single crops), spreading out across vast monocoloured landscapes, all lining up in their regimented rows. Nothing is really natural about these monochrome fields of gold, because the conditions these crops require to grow have never existed naturally. This is all land that has been flattened, drained, fertilised and smothered in pesticides to keep nature at bay. And if, just for a moment, we put all of those environmental issues to one side just for a minute, though, and look at the agricultural planet we have shaped, there is no doubt the achievement is truly remarkable. It's highly efficient crop production like this, which is producing enough food to cater for our ever-increasing human population and keeping many of those 8 billion (yes, 8 billion!) people out of starvation. But we of course can't forget the environmental issues that come with this miraculous feat. Forty per cent of the world's land surface is now given over to this kind of agriculture, to bland monocultures devoid of life. Of all the mammals today on Earth, only 4 per cent are wild animals, while 96 per cent are humans, their pets or domestic farm animals – a set of statistics that, from my perspective, is truly terrifying.

HUMAN PLANET

Above: *Earthrise* was taken from lunar orbit by astronaut William Anders in 1968, during the Apollo 8 mission. It was described as 'the most influential environmental photograph ever taken'. It highlighted the fragility and the transcendent beauty of our planet.

Overleaf: Signs of hope for the future: wildlife finds a way and animals are adapting and sometimes thriving even in the most urban spaces.

And so we arrive in modern times, the latest but certainly not the last chapter in a story that has been 4.5 billion years in the making. It's a story that has seen endless twists and turns; many of those moments we now deeply understand but much of our planet's history still remains hidden from our view. But the very fact we've been able to dig so far into the past, so deep into the life story of our planet, is a remarkable achievement and with it has come knowledge we simply cannot ignored.

It all began with an unremarkable planet, the third rock out from an unremarkable star, moulded from rocky debris. In its early years, Earth existed on the edge of annihilation. Battered by the violence of an unstable solar system and almost destroyed by the impact of another planet, it was a world with little evidence of what the future could hold. But then, from the thinnest of vapours, an atmosphere was formed, a protective shield beneath which the planet could provide the beginnings of stability against the hostility of space. And under this newly formed, reddening sky, rain fell and great pools of water collected into lakes and oceans, and it was somewhere in that water world, 4 billion years ago, that life somehow established its first foothold.

Ever since then life has been in an endlessly complex interplay with the Earth. At times the planet has looked as if it was conspiring to rid itself of this biological force; moments when it seemed that a frozen, choking or burning planet would return it to sterility, but it has never taken that final step and shaken off every last trace of life. At other times it's life that has seemingly had the upper hand, transforming the atmosphere and the Earth into a green and, at least for a while, pleasant land, when life was not just prolific, but wielded extraordinary power, only to witness once again how quickly millions of years of work can come undone. And now, one species, with exceptional talents and exceptional flaws, finds itself in exactly one of those moments, a moment where we hold immense power coupled with immense fragility.

This is the human planet, a world that may be awesome but for me cannot be described as wondrous anymore. Not when the planet is so scarred by our actions, riddled with the excesses of human consumption that continue to go unchecked despite the knowledge we now all collectively hold. In the last few days of filming for the television series we visited Mexico City, a vast metropolis of brick, concrete, steel and glass, not my sort of environment but an environment that cannot be ignored. Together with the twenty-one districts that radiate out across its 3,000 square miles, this is home to more than 21 million people and, staggeringly, that only makes it the sixth-biggest city on the planet by population. Across the world there are more than thirty other megacities, which are defined as cities that have more than 10 million inhabitants, and that number is only expected to grow. Now, I'm not saying for one second that cities are by definition bad things – they can, of course, bring huge benefits and efficiencies to how humans can live and work together. But as they grow bigger they make huge demands on resources that rely on our ability to produce food with brutal efficiency, to feed the millions of mouths. The foundation of all of these megacities, the ingredient that has powered all of this growth, is the energy that we have dragged out of the ground in the form of coal and oil and gas. Formed from life that died hundreds of millions of years ago, we have drilled and mined for these fossil fuels for over 200 years. Now, that's probably the smallest number I've written in this book and yet in the space of that infinitesimally tiny period of time, the Earth has been transformed. As we know only too well, not only has all of this fossil fuel powered the technology that's allowed us to build these megacities, but by pumping vast amounts of carbon dioxide into our atmosphere it has also started to radically transform our climate.

Above: The rise of megacities, like New Delhi in India, is changing the way we live, by supporting vast numbers of people in less space and requiring food and water to be piped in in massive quantities.

Of course, there is nothing new in life transforming Earth. In many ways, we are following in the footsteps of other organisms before us: like the bacteria that changed Earth's atmosphere to give us oxygen, or the swamp forests of the Carboniferous that created the coal that we now mine. But a key difference is that we are changing the planet and its climate with full knowledge and understanding of the consequences.

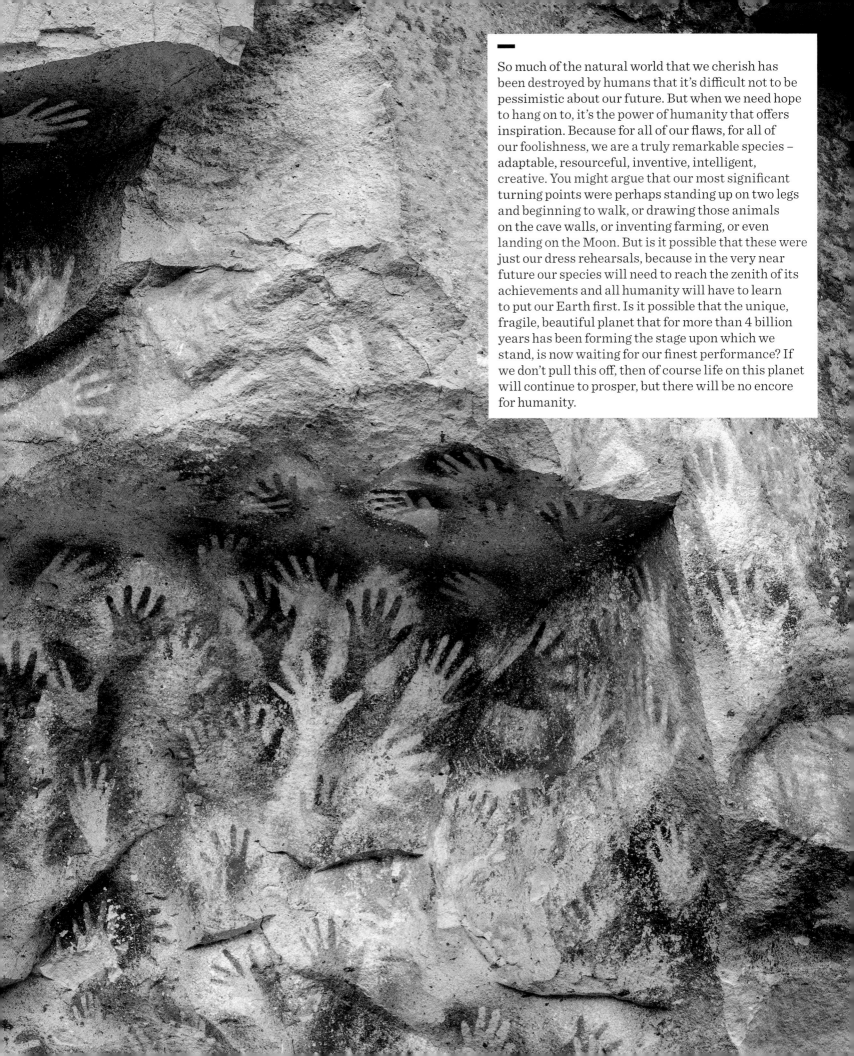

So much of the natural world that we cherish has been destroyed by humans that it's difficult not to be pessimistic about our future. But when we need hope to hang on to, it's the power of humanity that offers inspiration. Because for all of our flaws, for all of our foolishness, we are a truly remarkable species – adaptable, resourceful, inventive, intelligent, creative. You might argue that our most significant turning points were perhaps standing up on two legs and beginning to walk, or drawing those animals on the cave walls, or inventing farming, or even landing on the Moon. But is it possible that these were just our dress rehearsals, because in the very near future our species will need to reach the zenith of its achievements and all humanity will have to learn to put our Earth first. Is it possible that the unique, fragile, beautiful planet that for more than 4 billion years has been forming the stage upon which we stand, is now waiting for our finest performance? If we don't pull this off, then of course life on this planet will continue to prosper, but there will be no encore for humanity.

INDEX

PICTURE CREDITS

ii NASA

v Forgem/Shutterstock

vi S. ANDREWS (HARVARD-SMITHSONIAN CFA); B. SAXTON
(NRAO/AUI/NSF); ALMA/ESO/NAOJ/NRAO/SCIENCE PHOTO
LIBRARY

vii DETLEV VAN RAVENSWAAY/SCIENCE PHOTO LIBRARY

viii CORDELIA MOLLOY / SCIENCE PHOTO LIBRARY

x Franco Banfi/naturepl.com

3 Science Photo Library

4 John Cancalosi/Alamy Stock Photo

5 NASA/METI/AIST/Japan Space Systems, and U.S./Japan ASTER
Science Team

6 JACK DYKINGA/NATURE PICTURE LIBRARY/SCIENCE
PHOTO LIBRARY

8 AMERICAN PHILOSOPHICAL SOCIETY/SCIENCE PHOTO
LIBRARY

9 JIM REED/SCIENCE PHOTO LIBRARY

10 Eric Baccega/naturepl.com

11 Associated Press/Alamy Stock Photo

12 1 JOYCE PHOTOGRAPHICS/SCIENCE PHOTO LIBRARY, 2
PHIL DEGGINGER/SCIENCE PHOTO LIBRARY, 3 JOHN MEAD/
SCIENCE PHOTO LIBRARY, 4 DAVID R. FRAZIER/SCIENCE
PHOTO LIBRARY

13 1 PEKKA PARVIAINEN / SCIENCE PHOTO LIBRARY,
2 PEKKA PARVIAINEN/SCIENCE PHOTO LIBRARY, 3
PEKKA PARVIAINEN/SCIENCE PHOTO LIBRARY, 4 PEKKA
PARVIAINEN/SCIENCE PHOTO LIBRARY, 5 GEORGE POST/
SCIENCE PHOTO LIBRARY, 6 CORDELIA MOLLOY/SCIENCE
PHOTO LIBRARY

14 PEKKA PARVIAINEN/SCIENCE PHOTO LIBRARY

15 tl Pero Mihajlovic/Shutterstock, tr EUROPEAN SPACE AGENCY/
SCIENCE PHOTO LIBRARY, bNASA/VRS/SCIENCE PHOTO
LIBRARY

16 Ammit Jack/Shutterstock

17 t NASA/JPL, m G. BRAD LEWIS/SCIENCE PHOTO LIBRARY, b
John Cancalosi/Alamy Stock Photo

18 Andrew Mayovskyy/Alamy Stock Photo

20 NASA/GODDARD SPACE FLIGHT CENTER/SCIENCE PHOTO
LIBRARY

21 Ashley Cooper/naturepl.com

22 l NASA/Johns Hopkins University Applied Physics Laboratory/
Carnegie Institution of Washington, r NASA/JPL-Caltech

23 l US GEOLOGICAL SURVEY/SCIENCE PHOTO LIBRARY,
r NASA EARTH OBSERVATORY/NOAA/NESDIS/SCIENCE
PHOTO LIBRARY

24 Chris Packham

25 l THIERRY BERROD, MONA LISA PRODUCTION/SCIENCE
PHOTO LIBRARY, r NEW ZEALAND AMERICAN SUBMARINE
RING OF FIRE 2007 EXPLORATION, NOAA VENTS PROGRAM,
THE INSTITUTE OF GEOLOGICAL & NUCLEAR SCIENCES
AND NOAA-OE/SCIENCE PHOTO LIBRARY

26 robertharding/Alamy Stock Photo

27 t Chris Packham/BBC, b incamerastock/Alamy Stock Photo

28 l Historic Collection/Alamy Stock Photo, r TRphotos/Shutterstock

29 t NASA/JPL-CALTECH/UNIVERSITY OF ARIZONA/SCIENCE
PHOTO LIBRARY, b MAREK MIS/SCIENCE PHOTO LIBRARY

33 KARL GAFF/SCIENCE PHOTO LIBRARY

34 ml World History Archive/Alamy Stock Photo, b EYE OF
SCIENCE/SCIENCE PHOTO LIBRARY

35 t POWER AND SYRED/SCIENCE PHOTO LIBRARY, b Asahi
Shimbun/Getty Images

36 l EYE OF SCIENCE/SCIENCE PHOTO LIBRARY, r
seeshooteatrepeat/Shutterstock

38 stocksre/Shutterstock

39 l DIRK WIERSMA/SCIENCE PHOTO LIBRARY, r PannaPhoto/
Shutterstock

40 DIRK WIERSMA/SCIENCE PHOTO LIBRARY

41 Askar Karimullin/Alamy Stock Photo

42 t KARL GAFF/SCIENCE PHOTO LIBRARY, b MARK GARLICK/
SCIENCE PHOTO LIBRARY

43 INSTITUTE OF OCEANOGRAPHIC SCIENCES/NERC/
SCIENCE PHOTO LIBRARY

44 Andrea Pucci/Getty Images

45 Pierre Rochon photography/Alamy Stock Photo

46 Aaron Foster/Getty Images

47 NASA/SCIENCE PHOTO LIBRARY

49 Kertu/Shutterstock

50 t Stefan Christmann/naturepl.com, bl Ashley Cooper/naturepl.
com, br Bryan and Cherry Alexander/naturepl.com

51 t Ashley Cooper/naturepl.com, b Ole Jorgen Liodden/naturepl.com

52 Frans Lemmens/Getty Images

53 t Ed Reschke/Getty Images, b STEVE GSCHMEISSNER/
SCIENCE PHOTO LIBRARY

54 l Science Photo Library - STEVE GSCHMEISSNER/Getty Images,
r Iliescu Catalin/Alamy Stock Photo

55 © 2017 Bengtson et al, CC BY-SA 4.0 via Wikimedia Commons

56 Artem Gorlanov/Shutterstock, inset Chris Packham/BBC

58 Wild Wonders of Europe/Lundgre/naturepl.com

59 Olimpio Fantuz/4Corners

60 robertharding/Alamy Stock Photo

61 tl Valerio Ferraro/500px/Getty Images, tr Mark Chivers/Getty
Images, m Chris McLennan/Alamy Stock Photo, br Zmrzlinar/
Shutterstock

62 Design Pics/David Ponton/Getty Images

63 Susannah Porter

219 l PATRICK LANDMANN/SCIENCE PHOTO LIBRARY, 219 r
The Print Collector/Print Collector/Getty Images
220 Mark Newman/Getty Images
221 tl Sergey Uryadnikov/Shutterstock, tr Terry Whittaker/naturepl.
com, b Konrad Wothe/naturepl.com
222 PASCAL GOETGHELUCK / SCIENCE PHOTO LIBRARY
223 CNRI / SCIENCE PHOTO LIBRARY
224 t PROF. MATTHEW BENNETT, BOURNEMOUTH
UNIVERSITY / SCIENCE PHOTO LIBRARY, bl JORGE
GUERRERO/AFP/Getty Images, br LOOK AT SCIENCES/
SCIENCE PHOTO LIBRARY

225 robertharding/Alamy Stock Photo
226 Ongala/Shutterstock
227 t seng chye teo/Getty Images, b Martin Ruegner/Getty Images
229 t FLETCHER & BAYLIS/SCIENCE PHOTO LIBRARY, b oticki/
Shutterstock
230 NASA/SCIENCE PHOTO LIBRARY
231 Michele Falzone/Alamy Stock Photo
232 tl Sam Hobson/naturepl.com, tr Luke Massey/naturepl.com,
b Sam Hobson/naturepl.com
233 t Nayan Khanolkar/naturepl.com, b Doug Gimesy/naturepl.com
234 R.M. Nunes/Alamy Stock Photo

ACKNOWLEDGEMENTS

Telling the 4.5-billion-year biography of our planet is an ambitious starting point for any project, and back in the summer of 2021 we set ourselves quite a mountain to climb. Assembling a team that would be so brilliantly led by Rob Liddell and Raewyn Dickson, we began pulling together the very latest scientific research, to underpin the telling of five stories of the most dramatic moments in the Earth's history. The team from Moonraker VFX joined us to begin the complex task of accurately bringing so many lost worlds to life and our amazing production management team began navigating production plans that were still fraught with the complexity that the pandemic had brought to filming around the world.

Nothing would be straightforward on this project. As the series progressed through a succession of Covid disruptions, schedules needed to be ripped up, budgets reworked and editorial ambitions shifted again and again. And yet despite all these difficulties everyone always remained focused first and foremost on the wellbeing of the team and on the delivery of a world-class factual series. We are incredibly lucky to have had such a dedicated, skilled and talented team to navigate us through all of this and help deliver a project that fulfils all the ambition we set out with and so much more. This is just a small chance for us to say a very big thank you.

With Rob and Raewyn we had a leadership team that any production on the planet would be lucky to work with. Always calm in the face of adversity, kind in the running of the team and nurturing creativity every step of the way. Rob has led a team of producers and directors with a creative ease that hides the huge complexity he has dealt with so brilliantly over the last two years. So much of the success of this series is a result of his leadership and the creative vision he has shared so successfully with the team. Equally, Raewyn's skill in leading the brilliant production management of the project has allowed our ambition to remain undented whatever the world has thrown at us. Always calm, always ready with her wry sense of humour and always a delight to work with, thank you both for everything you have given to the project.

We have also been incredibly lucky to have a world class team of filmmakers working with us, with James Tovell, Duncan Singh, Cat Gale, Ben Wilson, Tom Hewitson and David Briggs producing and directing what are without doubt a beautiful series of films. We are very lucky to work with such dedicated and talented people.

They were supported by a hugely talented production team who have grappled with so many different challenges that have been overcome in endlessly creative ways. So a very big thank you to Jessica Springthorpe, Thomas Barnett-Welch, Grace Hartley, Kiki Lawrance, Jennifer Anafi-Acquah, Sophie Walsh, Emma Hyland, Sam Wigfield, Jay Balamurugan, Robbie Wojciechowski, William Hornbrook, Olivia Jani, Jeremy Parkinson, Warren Correia, Kate Jameson, Freddie Nottidge, Alistair Duncan, Imogen Ashford, Rebecca Rosenberg, Paul Saunderson, Professor Andrew Knoll, Ella Fitzhugh, Jemal Guerrero, Emma Hyland, Louise Salkow, Douglas Moxon, Ged Murphy, Bill Coates, John MacAvoy, Julius Brighton, Patrick Acum, Toby Wilkinson, Graham Boonzaaier, Mark Carroll, Randall Love, Nikki Bramley, Kendal Kempsey, Sebastian Blach, Zach Levi-Rodgers, Toby Wilkinson, Krish Thind, Marie O' Donnell, Vicky Edgar and the many, many other people who have supported this production. Including a big thank you to Simon Clarke and all the team at Moonraker VFX, all the team at Halo Post Production and our commissioning editors at the BBC, Jack Bootle and Tom Coveney, and all the team at PBS and WGBH.

Also a very big thank you to Nicola Cook, Tom Scott and Nik Sopwith for the brilliant development work that allowed us to make the *Earth* series. And as always, a very special thanks to Laura Davey for the incredible talent you bring to the running of the BBC Studios Science Unit.

Andrew would as always like to thank Anna, Benjamin, Martha and Theo for all the endless love that sits in our little corner of planet Earth.

Finally, a big thank you to the team at William Collins. Here we are, with the most beautiful book almost conjured out of thin air, it really is an utter pleasure to work with such talented people. So thank you to the brilliant Hazel Eriksson, Helena Caldon, David Price-Goodfellow and Chris Wright. And of course, a very big thank you to Myles Archibald – no dental emergencies this time just an appointment brought forward to add a little last-minute tension to the treatment schedule ... and endless reminders of the first (unbreakable!) rule of tie-in publishing.